A GUIDE TO THE

CORAL REEFS OF THE CARIBBEAN

MARK D. SPALDING

FOREWORD BY SYLVIA EARLE

CARTOGRAPHY BY GILLIAN BUNTING

UNIVERSITY OF CALIFORNIA PRESS
Berkeley Los Angeles London

Published by **The University of California Press**
University of California Press
Berkeley and Los Angeles, California
University of California Press, Ltd.
London, England

Clothbound edition ISBN: 0-520-24395-1
Paperback edition ISBN: 0-520-24405-2

*Cataloging-in Publication data is on file with
the Library of Congress.*

Citation: Spalding M.D. (2004). *A Guide to the Coral Reefs of
the Caribbean.* University of California Press, Berkeley, USA.

Cartography: Gillian Bunting

Origination: Swaingrove Imaging

Printed in: China

A **Banson** Production

Photos by the author and:
Alex Mustard: page 219
Callum Roberts and Julie Hawkins: pages 13; 17 (lower right); 19 (lower); 22; 23; 24;
35 (both); 36; 37 (both); 39; 90; 95; 97; 98; 136; 165 (lower); 173; 189; 205; 215; 221;
225 (right); 227 (upper left, lower 2nd right, lower right); 228 (upper left, upper 2nd
right, lower right); 229 (upper left, lower 2nd left); 230 (upper right); 231 (lower right);
232 (upper 2nd right); 238 (upper center); 239 (upper right, lower left)
Christopher Spalding: pages 28; 29; 187; 191
Colin Fairhurst: pages 52; 61; 86; 218
David Alonso Carvajal: page 203
Edmund Green: pages 19 (upper); 59; 79; 133; 181; 195; 230 (lower 2nd left);
231 (upper 2nd left); 232 (upper 2nd left); 237 (lower right); 239 (lower 2nd left)
Eric Riesch / marinelifeimages.com: page 231 (upper left)
Ernst von Weyhausen: pages 9; 93; 104; 107; 147; 155; 223; 230 (upper 2nd left);
236 (lower 2nd left); 237 (upper center, lower left)
Frank Borges Llosa / frankly.com photography: page 149
Mania Spalding: page 234 (lower left)
NASA: pages 10 (SeaWIFS Project); 47 (STS075-727-32); 99 (ISS003-E-7756); 173
(ISSOO4March20); 213 (STS077- 719-105)
Ned Deloach / marinelifeimages.com: pages 20 (lower); 124; 231 (lower left)
Neil Bourne: pages 89; 96 (upper); 112; 132
Paul Humann / marinelifeimages.com: pages 24; 25; 57; 80; 91; 148 (both);
159; 181; 217; 227 (upper 2nd left); 231 (upper 2nd right, upper right); 232
(lower right); 234 (lower right)
Peter Verhoog: page 105
UNEP / Denise Heitland / Topham: page 55
WWF-Canon / Peter CH Pritchard: page 129 (30090)
WWF-Canon / Sylvia Earle: page 142 (8429)
WWF-Canon / Ronald Petocz: page 240 (lower right) (2087)

FOREWORD
SYLVIA EARLE

In the words of Robert Breeden: "The Caribbean is not so much a place as it is a feeling." This echoes my own enchantment with that sapphire sea in the warm midsection of the Western Atlantic Ocean. I think of warm sea breezes on bare skin, the exhilaration of diving among pristine coral reefs glittering with fish, or the thrill of floating over seagrass meadows alive with dark shadows: a giant ray, a plump manatee, or a nurse shark on the prowl. For many, the "Caribbean feeling" is inspired by the splendid mix of cultures; the distinctive music; the tastes of spicy food, tropical fruits, and locally distilled rum. To others, it is the deep history – from geological roots, millions of years ago, to more recent centuries when pirates looted treasure ships that in turn had looted the New World. Today, millions of people regard the Caribbean itself to be the real treasure, a place to live or visit or dream about.

In *A Guide to the Coral Reefs of the Caribbean* Mark Spalding presents a fine, rich feeling for the natural, cultural, and historic character of the Caribbean. Here are illustrated, in microcosm, the urgent broad issues now facing the world's oceans and people, especially those whose lives are tightly connected to coral reefs. This is a book that will appeal to newcomers as well as seasoned residents of the region. With engaging, graceful prose, detailed maps, and a great array of images, Spalding provides an information-packed guide that is also an enduringly useful reference. At the same time, this is an appealing book just to be read and savored, either in small bites or in a single, non-stop feast.

Perusing this book, I get that Caribbean feeling: recalling the explosion of life gathered around lights at night; the silver sweep of tarpon racing after small fish; the flash of a squid turning pale, then iridescent, then dark before disappearing in a showy blast of molluscan muscle. In my mind I see the gray shadow of a whale shark as large as a city bus emerge, within a mass of spawning dog and cubera snapper. With a mouth as wide as the bumper of a minivan, and a tail taller than I am, she passes by, smoothly, intently, with deliberate grace, feeding on a profusion of fish eggs. I feel the warmth of sunlight penetrating through a glassy sea, sending shafts of light through a maze of branching coral, and watch silver bubbles of oxygen lift from the blades of nearby turtle grass. I see the reefs I knew as a child, the profusion of small fish, speckled lobsters, and sleek barracuda, the great Nassau grouper and the maze of trails in soft, marly sand made by countless large pink-lipped conchs. But I also see decades of change, of missing pieces, of barren seascapes, and then I long for my children and grandchildren, for all people everywhere to know what has been lost – and be inspired to do what it takes to restore health to the ocean.

Today, those visiting the Caribbean for the first time are impressed with the astonishingly clear waters, the unimaginably varied shadings of blue, emerald, aqua, and violet contrasting with white clouds and sand beaches, or equally stark black volcanic rocks and vividly green vegetation. At first glance underwater, the abundance and diversity of life is dazzling. Nearly all of the 30 or so major divisions or phyla of animals can be found on Caribbean reefs, whereas only about half of these occur in terrestrial and freshwater realms. On a single dive, a sharp-eyed swimmer might encounter starfish, urchins, sea cucumbers, crinoids. They will find corals, jellyfish, anemones, sponges, sea squirts, comb jellies, polychaete worms and squid – all are common underwater, but are found nowhere on land.

Those like Mark Spalding who have explored Caribbean reefs, and shallows and deep waters beyond over many years – myself included – will readily tell anyone who will listen: "We must take care to protect what we have, and restore what we can."

While still wondrously appealing, the lands

and surrounding waters of the Caribbean have in fact been shorn of much of their natural and cultural wealth. On a small island near the Dominican Republic, in 1494, Christopher Columbus and his crew encountered numerous sleek, furry creatures with the beguiling curiosity and the good nature inherent to most marine mammals. These were Caribbean monk seals. Thousands were killed for food and oil during the centuries that followed. The last ones were seen on the remote Serranilla Bank in 1952, and they are now officially listed as extinct.

Even by 1800, when world population hovered around 1 billion, before cars, planes, and electrical power – let alone computers, plastics, and space flights – humankind had set in train many profound changes. Vast tracts of forest and other natural lands were gone, and we were already embarked on diminishing the abundance and diversity of large sea creatures. The number of sea turtles had collapsed across the Caribbean, and the days when manatees grazed like great herds of cattle were over.

Swift population growth and the use of technological powers unimaginable even half a century ago have accelerated the changes. Recent studies have confirmed what fishermen already could sense: since the 1950s, 90 percent of the big fish in the oceans have been extracted. Blue fin tuna, swordfish, grouper, snapper, sharks and even the larger, colorful parrotfish and wrasses are mostly gone, and the giant pink conchs that once plowed through the sand flats and sea grass prairies have largely made their way into fritters, chowders, and ceviches. Our appetite for finding, catching, and consuming them has far exceeded their capacity to rebound.

But in the face of this change it is critical that we don't lose hope – the Caribbean still holds countless wonders. Wide expanses of reefs, seagrass meadows, and mangrove-laced shores remain. The marine life can endure, along with the people who rely on its benefits and love its beauty, but only if actions are taken immediately. In 50 years, up to half of the world's coral reefs have declined significantly or have been destroyed. But at least half still prosper. Even where 90 percent of the large fish are gone, 10 percent remain, and for these, unlike the monk seal, there is still time to reverse the decline.

Mark Spalding reminds us that in past centuries, no special care seemed necessary for the maintenance of the natural systems upon which all life depends. Air, water, and the fabric of living creatures worked together in a single system that included humankind. As the 20th century got underway, however, the need to protect areas on the land from human impacts became obvious. First in the USA and later worldwide, the national park and wildlife reserve concept grew within an overarching ethic of caring for the land. Sometimes described as the "best idea America ever had," national parks and other reserves have provided profound benefits to the economy, to human health, to our very survival. Could protection for the natural ocean systems and the historical and cultural wealth embodied there be less important? Might the 21st century mark the era in human history when an ocean ethic captures the minds and hearts of people and gives rise to a global system of significant marine protected areas?

Throughout this guide, there resounds a clear connection between ocean health and human prosperity, between protection of the natural assets and the benefits to humankind. With knowing, comes caring. As a valuable source of "knowing," this volume will help ensure that snappers will be spawning, whale sharks dining, coral reefs and people thriving in the blue Caribbean for many millennia to come.

Dr Sylvia Earle is one of the world's best-known marine scientists. At the age of 13 she first moved to the coast in Florida, and she never looked back. Her research career took her from Florida State University to Duke, to Harvard, then back to Florida. In 1970 she led the first all-female team on a research mission to live underwater for two weeks in a submerged research laboratory in the US Virgin Islands. She led pioneering work in the design and use of deep sea submersibles. In the early 1990s she was appointed Chief Scientist of the US National Oceanographic and Atmospheric Administration. Today, she is the Explorer in Residence at the National Geographic Society and the Executive Director for Global Marine Conservation at Conservation International.

ACKNOWLEDGMENTS

This book would not exist without the support and indeed the vision of its various sponsors. Thanks indeed to the Moore Family Foundation, the Henry Foundation, the Ocean Foundation, PADI Project AWARE, UNEP Caribbean Environment Programme, the World Wide Fund for Nature, and the UNEP World Conservation Monitoring Centre (UNEP-WCMC). Thanks also to the World Resources Institute, which generously made available its bathymetric data.

Gill's work on the maps has been phenomenal, but she and I would also both like to thank Corinna Ravilious for her advice all through, and for the years of work that she and others have put into building up UNEP-WCMC's wonderful map resources. Thanks also to Mark Collins, Ed Green, and Simon Blyth, all from UNEP-WCMC, for their input.

I would also like to thank all those whom I have dived with over the years, but especially the recent team in Cuba at Azulmar/Avalon, Joel and Filippo in particular. Also thanks to Stuart Cove and the team in the Bahamas, and the staff and team at Greenforce on Andros.

Amongst those who have commented on and guided the text I would especially like to thank my brother Christopher, my wife Mania, and my father John. Also critical for their input to specific sections were Wolf Krebs, Pedro Alcolado, Carmen Lacambra, Raymond Bideaux, David Buglass, and Tries Razak. The influence and inspiration of Colin Watkins is felt throughout. Indirectly the marine folks at Conservation International provided an important source of support through the work I did with them for the Defying Ocean's End conference. Many thanks to my co-author for that meeting, Phil Kramer, and to Linda Glover and her colleagues at CI. Thanks to Lauretta Burke and to the World Resources Institute for their work on *Caribbean Reefs at Risk* and their support of my involvement in this. Thanks also to the staff and resources of the Cambridge University Library map room and to Newmarket Library.

A number of excellent photographers have kindly provided an amazing array of photographs and many thanks indeed to Callum Roberts, Julie Hawkins, Ernst von Weyhausen, Ed Green, Colin Fairhurst, Neil Bourne, Christopher Spalding, David Alonso Carvajal, and Mania Spalding.

Finally I cannot thank enough Bart Ullstein for his work in bringing this together, and Helen de Mattos for her incredible skills in producing a book out of some very raw materials.

Supporting organizations
Moore Family Foundation
Henry Foundation
Ocean Foundation
PADI Project AWARE
UNEP Caribbean Environment Programme
World Wide Fund for Nature
UNEP World Conservation Monitoring Centre
World Resources Institute

CONTENTS

Using this book

We strongly encourage readers to get a feel for this book and get to know their way around it. Use the contents pages, flick through, and above all use the Index – if you read a species name somewhere and would like to know more, there may well be information, even a photograph, elsewhere in the book.

Key to maps

Bathymetry (meters)

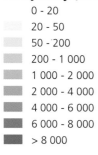

	0 - 20
	20 - 50
	50 - 200
	200 - 1 000
	1 000 - 2 000
	2 000 - 4 000
	4 000 - 6 000
	6 000 - 8 000
	> 8 000

Features

☐	Land	■○	Urban area/Town
■–	Coral	—	Road
	Mangrove	✈	Airport
	Lake	◣	Dive center
—	River	↙	Shipwreck
☐○	Protected area	▲	Mountain or volcano
-----	International boundary	▲	Underwater volcano

Note: Seas, oceans and other marine features are in blue type; terrestrial features are in black; protected areas are italicized in black.

Protected area abbreviations

AHI	Area of Historical Interest	MCD	Marine Conservation District	NWR	National Wildlife Refuge/Reserve
BiR	Biological Reserve	MFR	Managed Flora Reserve	PA	Protected Area
BR	Biosphere Reserve	MMA	Marine Management Area	Res.	Reserve
BR(N)	Biosphere Reserve (National)	MNP	Marine National Park	RS	Ramsar Site
ER	Ecological Reserve	MP	Marine Park	RZ	Replenishment Zone
FFPA	Flora and Fauna Protection Area	MR	Marine Reserve	RZSTP	Reserve Zone for Sea Turtle Protection
FFR	Flora and Fauna Refuge	NM	Natural Monument		
FFS	Fauna and Flora Sanctuary	NMP	National Marine Park	SPA	Sanctuary Preservation Area
GS	Game Sanctuary	NMS	National Marine Sanctuary	TNA	Touristic Natural Area
GSS	Grouper Spawning Site	NNP	Natural National Park	TMA	Tourism Management Area
IMA	Integrated Management Area	NP	National Park	WHS	World Heritage Site
LCA	Littoral Conservation Area	NR	Nature Reserve	WR	Wildlife Refuge
LP	Littoral Park	NRPA	Natural Resources Protection Area	WS	Wildlife Sanctuary

Map sources

The coral reefs, mangroves, and protected area information has been generously provided by the UNEP World Conservation Monitoring Centre. These are widely regarded as the "best available", but, as in any dataset of this nature, there are likely to be inaccuracies. Deeper "submerged" reefs are often not shown, though they may be described in the text, and in a few cases non-reefs may be incorrectly marked as reefs. The information comes from a wide range of different sources, most of which are listed in the *World Atlas of Coral Reefs*, though some have been updated.

The location of dive centers is based on original work conducted for this book, although based on a much smaller dataset originally compiled at UNEP-WCMC. Where multiple dive centers are very close together, they are usually represented by a single symbol.

The bathymetric data are largely based on a dataset prepared at the World Resources Institute for their *Caribbean Reefs at Risk* study – for a number of countries this data was improved using British Admiralty Charts.

INTRODUCTION

GETTING TO KNOW CORAL REEFS

Life began in the oceans. Here, over hundreds of millions of years, simple, minuscule organisms multiplied and diverged. Gradually, an array of creatures evolved, many quite unlike the animals and plants we know today. But amongst them were a few more familiar forms. About 500 million years ago a creature emerged which was simple, but in many ways quite beautiful. It had a short tubular body capped with a ring of tentacles that twisted gently in the currents of the ancient oceans picking out microscopic particles of food. This was the first coral.

By the time of the first dinosaurs, over 200 million years ago, this first coral had diverged into a multitude of forms. Some of these were the ancestors of modern corals, the creatures that lie at the heart of one of the world's most beautiful, most productive, and most diverse ecosystems.

Evolution

As they evolved, many corals developed the art of skeleton building. Some grew into vast colonies, laying down limestone skeletons like castles to house their delicate bodies. And where many corals lived together their combined skeletons, over the centuries, began to build vast structures These are coral reefs.

No-one who visits a coral reef today is left unchanged. These are nature's most spectacular ecosystems. Around the framework of corals thousands of other creatures live out their lives in a swirling complexity. Sponges stand erect or smother the rocks. The tall plumes of soft corals drift back and forth in the shifting waters. Lobsters and other crustaceans peer out from the deep recesses, and everywhere there are fish, darting and hovering, and sometimes shimmering in vast schools, impossible to count.

Take a dive

Picture the scene. You are in a small boat, rocking in the waves. The sun is hot and the glare on the ocean a little dazzling. A small island lies a mile or so off, but all else is blue: bright azure skies and the deep cobalt of the sea. Only the odd distant cloud breaks the spell. But you are ready to dive, with a heavy metal cylinder on your back, fins on your feet. Last minute checks, you pull on your mask, put a mouthpiece in your mouth and lean backwards to fall into the sea behind you.

Instantly your world is transformed and every sense is heightened. First your body becomes almost weightless – supported by water on all sides. The sea is a cool balm after the hot dry

French angelfish roam the reef in closely bonded pairs.

air above. You roll over and look around, your eyes adjusting to the gentler light, and a new world suddenly comes alive. It is not a tranquil scene, but one bursting with life. Up close, a yellowtail snapper swims in to look at you: a white fish with a smart yellow stripe along its flank and a yellow tail. Totally unafraid.

Below you is a busy sea floor. From above it all looks blue, with only the occasional flash of color from a darting fish. You drift slowly downwards and the dark shades of blue take shape, a forest of soft corals and the occasional towers of pillar corals rising up like fortresses.

Colors become discernible too and as you descend you realize that the ocean floor is packed with fish; a school of grunts is drifting amongst the soft corals. These sleek, streamlined, bright yellow fish rest in large numbers during the day, floating amongst the landscape of the reef. The corals are a multitude of soft shades: bronzes, ochres, tawny browns, yellows, greens, blues, and pinks. Sponges bring flashes of red and purple, yellow, and almost iridescent azure. Other tiny fish flash like brilliant jewels – these are fairy basslets, and they disappear under a small overhang as you approach.

Viewed from space, shallow waters lie in a marvelous palette of bright colors around the coasts and islands of Florida, the Bahamas, and Cuba. Coral reefs fringe the edges of these turquoise seas, tumbling into the deep ocean all around.

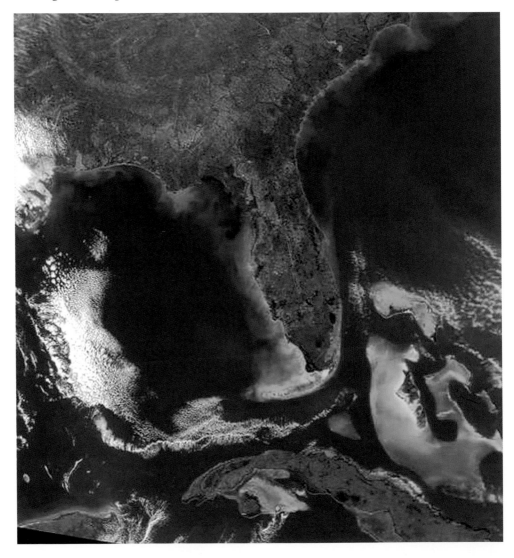

From here on, every dive is different. One day you may be treated to a fly-past by a group of majestic eagle rays. Another time you may spot the almost invisible form of a scorpionfish, or the bizarrely-shaped slipper lobster, like an armored vehicle. You may be engulfed by a swirling school of jacks, appearing out of the blue like a wall of silver. In shallower water you may find fabulous gardens of elkhorn or staghorn coral which grow where wave action is stronger, throwing up wild, jagged arms in defiance of the shifting waters all around.

Critical resources

Coral reefs are not widespread in the world. They occupy less than one thousandth of the area of the oceans, but this small area contains about a quarter of known marine life. They thrive in warm clear waters where there is little pollution. On a world map, these conditions translate into a broad swathe around the tropics. The Caribbean, with its long continental coasts and broad scattering of islands, is one of the great centers of coral reefs.

Long isolated from the coral reefs of the Pacific and Indian Oceans, life in these waters has diverged, and almost all of the species found here – sponges, corals, crabs, urchins, lobsters, fish – are unique to the Caribbean.

Humans first came to these shores about 11 000 years ago, and reached at least some of the islands by 6000 BC. These early inhabitants made considerable use of the natural resources of the oceans, fishing and hunting turtles, as well as gathering shellfish. The arrival of Europeans wrought massive changes, firstly with the demise of millions of native Americans, but later, as the new occupiers began to increase in number, on the natural ecosystems themselves.

There is no doubt that the coral reefs of the Caribbean just 500 years ago were even more spectacular than they are today. At this time the Caribbean monk seal was found lounging on the beaches. Turtles were so numerous in places that boats would be constantly bumping into them. Such large creatures were quickly hunted out, or became very rare. In the 20th century there was even more damage, with overfishing and the loss of entire reefs due to pollution, disease, and smothering by sediments from the land.

Even so, much remains. The blue Caribbean still teems with life. People across the region rely on the coral reefs for food, and the coral barriers protect many coastlines from the worst ravages of storms.

Scientists began to document the underwater world some centuries ago; however, their efforts were rarely based on direct observation. The sea floor remained hidden from view and few people had any idea of its vibrant beauty. This all changed in the middle of the 20th century, when Jacques Cousteau and others developed the Self Contained Underwater Breathing Apparatus (SCUBA). The ability to breathe underwater transformed our view of the world. Within just a few years diving became a popular pastime and, in 1957, a diving club opened in Jamaica – probably the first coral reef diving club in existence. A door was opened.

Now it seems hard to believe, but the magical underwater world of the coral reef was invisible to us just 60 years ago. Today there are well over 15 million recreational divers. Many of these have thrilled to dive on a coral reef, and their interest has sparked real concerns and growing efforts to bring back the reefs, to allow them to recuperate and become healthy, productive habitats once again.

This book provides a toolkit. It allows those who want to learn more to become expert in the world of corals and fishes. It provides advice to those who care, letting them know where to go and what to do. It is intended to enthuse and enthrall.

There is no better way to enjoy a visit to a coral reef than to understand it. As one learns the names of the fish and the corals they become more familiar. When one can see a tiny fish swim into the mouth of a giant barracuda, and then watch it swim out again it is truly marvelous to know that the giant was actually having its teeth cleaned. To see a coral is wonderful, but to know how old it is, and how it feeds, gives one a new perspective. And central to the book is a guide to the region, a guide to what happens where, and what one can expect to see in every country across the Caribbean.

Armed with this knowledge we can enjoy coral reefs all the more. And by visiting the reefs and enjoying their beauty we can contribute to their protection.

1.1 THE WORLD OF REEFS

WHAT ARE CORAL REEFS?

Corals are a group of animals found in all the seas of the world, from deep Arctic waters to desert shorelines. They are best known from warm tropical waters, where they are found in a fabulous array of forms. Soft corals, shaped like bushes, feathers, or fans, waft in the ocean currents, while stony corals are solid and unbending, sculpted into a multitude of shapes.

The coral animal

With such variety it becomes difficult to understand what defines a coral. In fact to answer this we have to take a close-up look. Most of the corals in the Caribbean are actually colonies. They are built by thousands of individual animals known as coral polyps. Each polyp has a simple, tube-shaped body, with a circular mouth at the top which also serves as an anus. Around the mouth are numerous tentacles which help to capture food. Inside the mouth there is a throat or pharynx and then just a single large cavity or gut called a coelenteron, which has a number of flaps or mesenteries poking into it.

There is no head or brain, but corals have a simple nervous system, called a nerve net, which enables them to control the muscles in their body walls and especially in their tentacles. The tentacles can also sting. They have minuscule capsules (nematocysts) on their outer surfaces, which contain an inverted whip, tipped with poisonous spines. These can be fired with explosive force: the whip lashes out and impales the victim, usually a small or microscopic animal in the water.

Most coral polyps can create new coral polyps through a process known as budding. This happens when an existing polyp splits in two, or grows a new polyp out of the side of its body. The new polyp remains attached to the old one and in time may also bud. Soon there may be hundreds of identical polyps all joined together in a colony of genetically identical clones. Most of the corals that do this have developed another trick. They build simple skeletons to support their combined bodies.

These great colonies are the most familiar forms of the animals we call corals, and it is their skeletons which help to build and define coral reefs. Living and working together the polyps have created a sort of "super-animal", but the polyps themselves remain the distinct units, the building bricks in the reef architecture.

How a reef is built

Coral colonies build a great variety of skeletons. Some rely on little more than the pressure of water within their tissues (hydrostatic pressure) to hold themselves upright. Others use stiff proteins, building whips, nets, or branching trunks which are still flexible in the moving water. Most important, however, in the making of coral reefs is a group known as stony corals. These lay down mineral skeletons from calcium carbonate, or limestone. Like the soft corals, they come in a fantastic array of shapes – some encrusting the contours of the sea floor, others rising up in great domes, tangled branches, tall

Figure 1: Cross-section of a polyp. The skeleton is shown in yellow.

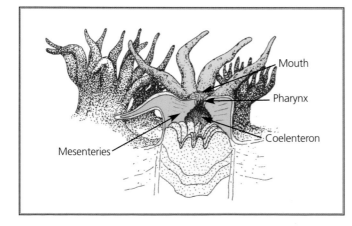

Mouth

Pharynx

Coelenteron

Mesenteries

towers, or delicate scrolling leaves. It is their work which has created the world of coral reefs.

Stony corals grow quite slowly – the domed shape of massive corals may only grow outwards by a centimeter or two each year, and even the fastest growing branching forms may extend their branch tips by only about 15 centimeters per year. Year upon year, however, these corals lay down huge rocky skeletons reaching several meters across. When a coral dies, others grow in its place, building upwards on the older skeleton. During storms many corals can be smashed to rubble by the waves, but the pieces settle into the landscape, filling the gaps between the surviving corals. Some types of algae also lay down limestone skeletons, and some of these provide a kind of cement, filling the spaces and helping to bind together the coral and rubble. So gradually, year on year, the sea bed becomes covered in coral rock and begins to extend upwards. Corals and algae grow on the remains of their ancestors, and a reef is born.

A coral reef, then, is not like a field or a forest on land, but more like a mountain range. It is a physical structure, built over hundreds of years by corals, and still covered with a living layer of corals and other life. It is a huge, three-dimensional seascape, rising up from the sea floor towards the ocean surface. From land, or even from space, these strange living mountains are clearly visible as patches of bright blues and greens – vivid against the dark blue of the surrounding deep water.

The most actively growing part of a coral reef may grow upwards by 4 meters in 1 000 years, so it can take many thousands of years for really big reefs to be built. Most of the largest reefs in the world have grown in phases – periods of reef growth have been halted by changes in climate or sea level, then begun again as conditions ameliorated. The latest period of reef growth began at the end of the last ice age, only 10 000-12 000 years ago, and in fact many Caribbean reefs did not start to grow until about 6 000-9 000 years ago. Scientists have drilled through the outer slopes of the Belize Barrier Reef and found that about 16-18 meters of coral materials have been laid down in the last 7 000 years. But this reef, like many others, is growing on foundations laid down by a much older coral reef in a different geological era.

Creole wrasse on a reef wall with sponges and fan corals.

MEASURING REEFS

Measuring the size of a coral reef is not a simple task. Most of them are broken up by channels or areas of deeper water, and so it is difficult to say where one reef ends and the next begins. A few very large structures, such as the barrier reefs of Belize, Florida, and Cuba, are nearly continuous, and provide rival claims as the largest single reefs. At the same time, however, scientists point out the high levels of connection even between reefs quite far apart.

These wider "reef systems" can be very large indeed. In Central America, the Mesoamerican Reef is currently receiving a great deal of attention. Scientists consider that the extensive reefs running from Mexico to Honduras may, to some degree, be "self-contained", being both sufficiently large, and sufficiently isolated from other areas, to function as a single system. More work is needed, but it seems likely that equally large, self-contained systems may exist, for example, around the extensive coastline of Cuba, or across the Bahama Banks.

To keep such comparisons in perspective, it is worth pointing out that Australia's Great Barrier Reef is actually larger than all of the Caribbean coral reefs combined. It is not the size of the reefs that is important, but the diversity, the complexity, the productivity, and of course their interest and value to people.

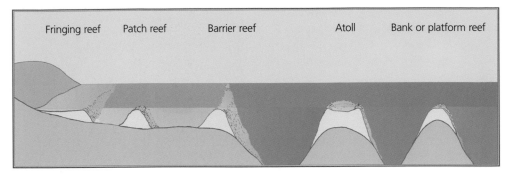

Fringing reef   Patch reef   Barrier reef   Atoll   Bank or platform reef

Figure 2: Various types of coral reef. Note that bank barriers, which are common in the Caribbean, are similar to barrier reefs but separated from the shore by shallow water rather than a deep lagoon.

Living with algae

Almost all the corals on a coral reef live in a remarkable partnership. Within their tissues lives a very simple type of algae, known as zooxanthellae (pronounced zoo-zan-thel-ee). These live in quite extraordinary numbers – a cross-section through a polyp would reveal 1-5 million zooxanthellae in every square centimeter. Like all algae, the zooxanthellae use special pigments to capture the energy from sunlight. In the process known as photosynthesis they use this energy to convert carbon dioxide and water into sugars, which they need in order to grow.

The coral polyps provide a safe and secure home for the zooxanthellae, and in return the algae provide the corals with sugars and other food which help the corals to grow. Thanks to this partnership, corals can thrive even in places where there is little food to be filtered from the water. Many reef corals get 80-90 percent of their food from these tiny algal partners and catch very little for themselves. It is the photosynthetic pigments of the algae which give the corals their beautiful colors. And it is this dependence on sunlight which means that reef corals only thrive in clear shallow waters.

Types of coral reef

Charles Darwin was fascinated by coral reefs and was one of the first to build a clear picture of how they might be formed. He saw how stony corals could lay down solid skeletons and how they needed sunlit waters. He saw how vast edifices in the oceans could have been built, over millennia, by the simple labors of stony corals. Darwin went on to describe three "classic" coral reef structures, and to consider how they might be formed: fringing reefs, barrier reefs, and atolls. These terms are still used to describe many coral reefs today. At the same time, other terms have

Figure 3: The typical reef zones of a barrier reef, many of which are also found in other reef types. Photos of the different zones are provided overleaf.

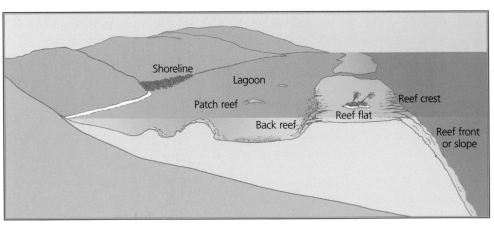

Shoreline

Lagoon

Patch reef

Back reef

Reef flat

Reef crest

Reef front or slope

also been devised, particularly in the Caribbean, to describe other coral reef formations that do not fit so readily into Darwin's model.

Fringing reefs

Perhaps the commonest reefs of all, found along the coast where the corals have built up shallow platforms in the brightest waters. The outer edge of this reef fringe is often a slightly steeper slope where the most active coral growth occurs.

Barrier reefs

Long reefs lying offshore, separated from the land by a slightly deeper lagoon. They are usually on the edge of the continental shelf facing deep water. Darwin suggested that they might have been formed by the upwards growth of a fringing reef when the land began to sink, or the sea level began to rise.

Coral atolls

Ring-shaped reefs surrounding a lagoon, found rising up from deep oceanic water. Darwin suggested that they might have formed when volcanic islands sank down, over geological timescales. The growth of the corals, he suggested, enabled them to keep up with the sea surface and the reefs remained after the island had disappeared.

Bank barrier reefs

Similar to barrier reefs, these are long reefs lying offshore, but lack the deep lagoon running between reef and shore. These are more common than true barrier reefs in the Caribbean, and form some the largest reef systems.

Bank or platform reefs

Like atolls, these typically rise up from deep oceanic waters, but lack clear circular structures. In the Caribbean they are again far more common than true atolls.

Patch reefs

Small outcrops of reef in shallower water, sometimes isolated, sometimes in an area where fringing or barrier reefs are also found.

Many of these structures do not rise up close to the water surface (this is a particularly common feature of Caribbean coral reefs) and the term

FADING LIGHT

One of the major drivers of patterns of life on the reef slope is the amount of light filtering through the water. When sunlight hits the sea surface, some is immediately reflected, giving us the dazzle off the water surface, but most gets through to light up the bright surroundings of the shallow reef. Going deeper, the water absorbs different wavelengths of light at different rates. The reds are one of the first colors to be blocked out, often in the first few meters from the surface, while yellows and oranges go soon after. Violets and greens will penetrate to a depth of 50 meters in clear water, but blues and greys penetrate deepest of all. In the clearest oceanic waters these last vestiges of light may penetrate the sea to depths of 200 meters.

Corals need sunlight to power their algal partners. In the very clearest waters, stony corals have been found in depths of just over 100 meters, but more typically, few thrive below 50 meters and many are only found in much shallower waters. Fish and other animals make use of the sunlight too. Unlike other marine ecosystems, colors are constantly used for signaling as well as for camouflage. Looking at pictures in a book taken with artificial light, it is easy to forget how things really look underwater. Many creatures on the reef are colored red and yellow, but in the filtered light below 10 or 20 meters, these organisms will not appear bright at all, and will blend into the shifting shadows of the reef slope.

The sea water quickly filters out light. At 20 meters and without the help of a powerful light, it seems hard to understand how the red hind gets its name.

submerged is quite often used, in conjunction with the above reef categories, to describe reefs that are 5 meters or more below the surface.

Zones of the reef

Just as different reefs are defined by their overall shape or structure, so each reef structure can be clearly divided into different zones. Each zone has very different characteristics – brighter or darker, warmer or colder, more or less exposed to waves, or to the influence of sediments or freshwaters coming from the land. As a result of these different conditions, each zone is home to numerous creatures specially adapted to live there and nowhere else.

Where the reef has grown up close to the sea surface to form a shallow platform, this is known as the reef flat. This can be a place of harsh temperatures and, if there is land nearby, mud and freshwater may wash into these shallow waters. It is a difficult world for many species, and, although it was built by corals, only a few survive here.

The outer edge of this reef flat is known as the reef crest – here the shallow water is constantly mixed with offshore waters, so that temperatures are more stable and the water much clearer. In some places, however, powerful waves crash on to the reef crest. It can be a vibrant area, buzzing with life, but in more exposed places only a few species survive.

The deepening waters below the reef crest are many people's image of the classic coral reef. Conditions are near perfect, with bright, clear, well oxygenated waters, and there is a constant jostling for space on the sea floor. This area is known as the reef front or reef slope. It may be gently sloping, or fall quite steeply into deeper waters, but as the water deepens another change is wrought. The water itself begins to filter out the sunlight, which is the driving force for all life on the reef, and slowly the abundant life begins to decline.

Figure 3 illustrates the main zones across a barrier reef, and photos of these zones and some of their characteristic features are provided opposite. At least some of these same zones are found on most reef types. Smaller, submerged reefs, however, are not really zoned, and here the entire reef is similar to the busy, thriving world of the reef front.

1. Shoreline. *This may be sandy, rocky, or even muddy. In many parts of the Caribbean there are mangrove forests fringing the shore, blurring the boundary between the land and the sea.*

4. Back reef. *The area on the short slope from the lagoon up to the reef flat. Rather like patch reefs, the shelter provided by the main reef beyond means that large areas of delicate and complex corals can flourish.*

7. Reef front, reef slope, or fore reef. *Beyond the reef crest the coral reef slopes down into deep water. This is where the greatest concentrations of life are found. Different species are located at different depths, with elkhorn and staghorn corals giving way to soft and pillar corals, and a range of brain and star corals.*

2. **Reef lagoon.** *A body of water, usually reaching over 2-3 meters deep, which is enclosed between the land and the coral reef offshore. The lagoon waters can be muddy or clear, and often support extensive areas of seagrass beds.*

3. **Patch reefs.** *These often rise up in clearer lagoon waters, where calm conditions can allow the growth of large and intricate coral formations. Immature fish often spend time on patch reefs before moving out to the offshore reefs.*

5. **Reef flat.** *A shallow band running from the shore in the case of fringing reefs, or making up the top of the barrier reef or atoll rim. Built by corals in times past, it is dominated by bare rock, sand patches, and perhaps seagrasses; sometimes a few corals are found.*

6. **Reef crest.** *The shallowest part of the reef, often marked by breaking waves. In some reefs, elkhorn, thin leaf lettuce, and fire corals survive here, but if wave action is stronger the crest is dominated by coralline algae or hardy species such as zoanthids.*

8. **Spur and groove.** *On many reefs where there is moderate to strong wave action, the shallower parts of the reef slope may be deeply incised with channels (grooves), running between high buttresses (spurs). The bottom of the grooves tends to fill with sand or rubble, but the edges of the spurs are often richly adorned with life.*

9. **Reef wall.** *Though not strictly a reef feature, in many coral reefs across the Caribbean the reef slope ends quite abruptly with a very steep to vertical wall of rock, falling to depths of 100 meters or more. Such walls tend to attract particular life forms, including deepwater sea fans, rope sponges, black corals, and sheet corals.*

LIFE ON THE REEFS

Coral reefs are the most complex and diverse ecosystems on Earth. Nearly 100 000 different types of coral reef plants and animals have been described by scientists, but this is only scratching the surface. If we could ever finish the count, there would probably be between 1 million and 3 million different coral reef species.

At the level of an individual reef, the diversity is bewildering – one Caribbean study found 534 different species from a single plot of only 5 square meters. In addition, these represented 27 different major groupings, or phyla – each an utterly different life form. Some of these life forms, and the major species found on the reef, are described in the final chapter of this book.

One of the most characteristic features of reefs, like other diverse communities, is the incredible variety of interactions. Each species occupies a unique role, or niche, on the reef, but with so many species there is enormous pressure from all sides to change and adapt: to find a space to live and grow, to eat the uneatable, to avoid being eaten, to hide, to hunt, and to reproduce. With 500 million years of evolution, the coral reef represents a treasure house of ecological invention. Here we consider just some of the workings of life on the reef.

Eating and being eaten

Plants lie at the base of almost every ecosystem. Converting the sun's energy into food, they provide sustenance not only for themselves, but also for the great chains of life which feed on them, and on each other. Adjacent to many coral reefs, the worlds of mangrove forests and seagrass beds are home to plants that are quite familiar – rich, green, and clearly edible (at least to some creatures). But on the coral reef itself the plants are less obvious. Simple algae cover the bare surfaces with a fuzz of brown or green, or grow into the pretty forms of various "macroalgae". Others drift unseen in the plankton. Zooxanthellae are an algal form which lie hidden within the tissues of corals, invisible, but powering the animals which build the coral reefs.

Grazing and gardening

A host of mouths scrape and chomp at the surfaces of the coral reef. The celebrated queen conch grazes on seagrasses and algae in shallow waters, and a few other snails do the same. Some of the most important grazers are sea urchins, but perhaps the most familiar are a host of fish. Some, such as parrotfish, rove widely over the reefscape, diving down occasionally to scrape and pick at algae. Several of the diminutive damselfish are known as farmer fish. They protect a tiny patch of reef, chasing off all other grazers so that the algae can grow into thick mats for their personal consumption.

Drifting food

Drifting in the sea, many plants are so small as to be totally invisible. These form part of the plankton, the floating ecosystem that drifts throughout the world's oceans. Unseen struggles for life are played out here – the algal cells, or phytoplankton, are eaten by tiny animals (zooplankton), which in turn are eaten by slightly bigger animals.

Many of the larger creatures of the reef feed on this plankton. Sponges are some of the

Plankton swarms around this pillar coral at night. The coral's tentacles catch and hold many of the smaller creatures of the plankton.

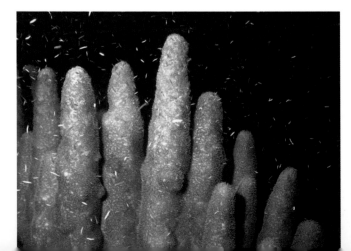

simplest: they draw in water through tiny holes, or pores, on their surfaces, and use filters to pick out food before the water leaves through larger pores. Feather duster and Christmas tree worms have far more fabulous structures to filter the water – poking brightly colored feathery heads out of the corals and sand. At night feather stars and basket stars clamber up the reef to prominent outcrops where they wave long feathered or branched arms out in the water to see what they can catch.

Corals, and their relatives such as sea anemones and zoanthids, also feed on the plankton, and some take larger prey, including animals just about visible to the naked eye. Within their tentacles, nematocysts carry a powerful poison-tipped barb to hold and immobilize their prey, which they pass to their mouths for digestion.

Countless fish also feed on the plankton. Many of the shimmering schools of brown chromis, sergeant majors, and even the swirling crowds of creole wrasse, can be seen picking at apparently invisible particles in the water. Perhaps strangest of all are the great giants of the fish world – the fabulous manta ray and the gigantic whale shark are both gentle filter feeders. Swimming along with their mouths agape they scoop up vast quantities of water which they filter in a constant flow through their gills.

Attack and defense

In nature the pressures to feed, and to avoid being eaten, are of paramount importance, and in the maelstrom of life on the reef these pressures have led to extraordinary adaptations. Through evolution, one species may develop a strategy to avoid its predators, but at the same time those same predators must evolve a strategy to outwit their prey. An arms race ensues, with species locked into ongoing battles of tit-for-tat adaptation.

Chemical warfare

Sponges are among the reef's most primitive creatures. There can be few simpler animals, and at first glance they look quite defenseless, sprawled out in the open and unable to move. However, they have evolved complex chemical compounds which poison any creature that tries to eat them. But the hawksbill turtle goes out of its way to feed on sponges. This ancient species

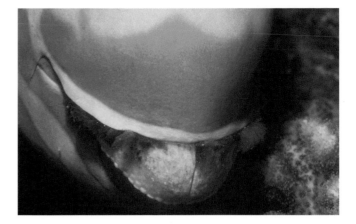

Parrotfish have highly adapted teeth, forming a beak with which they scrape the seabed, and even break off chunks of rock in order to get at algae.

Several algal-feeding damselfish, such as this threespot damsel, fiercely defend small patches of the reef from other grazers, allowing a farm of rich algae to grow.

has a powerful digestive system which can break down some of the toxins, but it has also adapted its behavior, and will only take small amounts of each sponge at a time, to avoid ingesting a surfeit of any one toxin.

The shell-less snails known as nudibranchs are a fascinating group – all are specialized feeders, many feeding on sponges, others on hydroids (relatives of the fire corals). Many are highly poisonous, and advertise the fact by their flamboyant colors, but it is the source of these poisons which is so extraordinary. The nudibranchs ingest the poisons of their hosts, but do not digest them. Rather they hijack them, storing them up in their body tissues so that they become poisonous to others. The species that eat

A bar jack shadowing a Spanish hogfish. Sticking close to a non-threatening fish, the bar jack hopes to sneak up on potential prey.

Decorator crabs such as this neck crab adorn their bodies with other animals, such as hydroids, as a form of camouflage.

hydroids are even able to hold on to the stinging organs (nematocysts) from their prey, storing them like loaded weapons, which they can fire off when threatened.

Hide and seek

One of the easiest ways of avoiding being eaten is to avoid being seen, or heard, or smelt. This same tactic is used in reverse by predators – invisibility can mean that your prey comes to you, with no further effort on your part.

Hiding is a common approach. Sea urchins, shrimp, brittle stars, and many different mollusks rest out the day in the myriad holes and tunnels of the reef, but at dawn and dusk there is a great change-over. The nocturnal invertebrates move out to feed, while countless fish take up the same hiding places in the reef framework. Other diurnal species make sure they have a hole to retreat to even during the day – yellowhead jawfish build quite elaborate burrows; garden eels

never quite leave the safety of their holes, keeping their tails firmly in the sand at all times, while the razorfish with their strange blunt heads make a nose-dive into the sand whenever they feel threatened.

Camouflage is another common tool. Few could make a more concerted effort than the decorator crabs, which create their own costumes out of other reef life. They use their pincers to pick up pieces of reef life, such as sponges, algae, or hydroids, and stick them to their shells and their legs, so that the individual crab becomes almost invisible.

Sit-and-wait predators, notably fishes such as scorpionfish, toadfish, pancake fish, and the greatest masters of them all, the frogfish (see page 181), have adaptive camouflage that uses color, shape, and texture to ensure that they blend in with their surroundings. The fleshy growths on their skin look so like algae that many divers just assume that's all there is.

A great many species on the reef can change color. Frogfish do this remarkably well, but for them the process is slow. Another sit-and-wait predator, the flounder, is far more dramatic, and can change color to match its background within seconds. These fish, along with cephalopods (octopus and squid), achieve these changes using special cells in their skin known as chromatophores which contain colored pigments – they are able to concentrate these pigments into a small space (essentially making them invisible) or spread them out through the cell to create different effects.

Other predators choose stealth to approach their prey. Quite a number of fish practice the art of shadowing. This is where a predator, such as a trumpetfish or a bar jack, tucks in close beside or behind another similar sized or larger fish and uses it as cover – hoping that its prey will not spot its approach.

Physical defense

Many organisms use simple physical armor to avoid being eaten. Most mollusks have thick shells; sea urchins and lobsters use a combination of hard carapaces and sharp spines. A few fish have also taken the armored approach, such as the armor-plated bodies of the boxfishes. Others change size to put off predators. The pufferfish fills its body with water when threatened, becoming too large to swallow. The porcupinefish relies on both armor and enlargement – sharp spines spike outwards when the fish is inflated.

Safety in numbers

Many fish in the world of the coral reef are found in schools, which can number from a few tens to hundreds, even many thousands of fish. Schooling occurs for a number of reasons, mostly related to attack or defense. For smaller fish the school represents a large but confusing target for a predator. For the predator it is not enough to target a school: it must single out an individual fish if it is to make a catch, but in the swirling confusion of many fish this can actually be very difficult. The silvery colors adopted by fish such as silversides serve to heighten this confusion – as the fish swerve and dart, the reflections further confuse their predators.

Other fish live in schools to help them find food. In the case of surgeonfish, a dense school is enough to overpower the efforts of territorial damselfish defending their algal farms, or even their nests. The latter may nip wildly at one or two individuals in the school, but are quickly overwhelmed. Some of the fish eaters choose to hunt in schools. Feeding jacks have been observed to work as a team in trying to feed on schools of silversides. They race repeatedly through the school, and manage to break off smaller groups, from which it is much easier to pick out their prey.

Reproduction

One of the commonest modes of reproduction on the reef, as described for the corals (see box), is known as broadcast spawning. Many creatures, but especially those which cannot move such as sponges and tubeworms, or which do not move far or fast such as many echinoderms, simply broadcast their eggs and sperm directly into the water, hoping that some may mingle with others of the same species.

Clearly this is a hit-and-miss strategy, and most broadcast spawners will release many millions of gametes at a time. Like corals they use a series of environmental cues to ensure they spawn in synchrony. These cues are often the same for many species. During the spawning season it is quite common to come across quite a range of creatures releasing small wafts or thick

clouds of gametes. In Bonaire, for example, the long-spined sea urchins typically spawn around noon, the touch-me-not sponges during the afternoon, Christmas tree worms at around 7.00 pm, and the brittle stars, often gathered into small groups and clambering on top of one another, between 7.30 and 8.30 pm. Few can be more impressive than the giant barrel sponges, which, for just a few hours each year, become like huge smoking chimneys, with clouds or thick clots of gametes drifting out of their cavernous chambers.

Many reef fish also rely on the rather haphazard release of gametes into the water, but, being more mobile, they come together to ensure

CORAL REPRODUCTION

Budding is a highly effective way of getting bigger as a colony, but as coral polyps are unable to move around, they also need a way of setting up new colonies. To do this they produce tiny oval-shaped larvae known as planulae which live in the plankton for a few days or weeks. They are carried by currents, but are also capable of a limited form of swimming. When they are swept close to the sea floor, if the conditions are right, they swim down and settle on the bottom to form a polyp, and to begin a new colony.

This much is simple, but for many years scientists were very sketchy on the details. They assumed that most corals released already formed larvae into the water, but in the early 1980s, a group of Australian scientists uncovered what must be one of the coral reefs' most spectacular events. Over just a few nights, once a year, they found corals were releasing not larvae, but eggs and sperm into the water in great quantities. And they found that most of the corals right along Australia's Great Barrier Reef were doing this at the same time.

Since that first discovery, scientists have identified other dramatic coral spawning events all around the world. Corals spawn at night and the orchestration of this event is quite phenomenal. Linked to a series of cues, including the moon, the water temperature, the timing of sunset, and certain chemicals in the water, all the corals of a particular species, even if they are miles apart, release their spawn within minutes of one another.

Dive centers in many countries are now taking divers and snorkelers to witness this extraordinary event first hand (see page 219). In most corals the spawn is released in little balls of orange, pink, or white. They emerge in waves forming great clouds as they drift slowly towards the sea surface. Rather like an inverse snowstorm, divers find themselves surrounded by gently eddying waters filled with tiny balls. At the surface they burst open, releasing milky clouds of eggs and sperm. The water becomes cloudy and, with such vast quantities of material being released into the plankton, plankton-feeding animals are overwhelmed, giving at least some of the young planulae a chance of survival.

With an awesome array of spines the web burrfish already looks unpalatable, but when threatened will puff itself up with water, making it too large for many fish to swallow.

a more direct mixing of eggs and sperm. For some species, such as the abundant bluehead wrasse, sex is an almost daily occurrence. Males of many species, such as the queen parrotfish, maintain territories which might encompass groups, or harems, of females. In these situations they spawn with several females together, or in sequence.

There is very little fidelity on the reef, but some fish only spawn in pairs, and a few species, such as the French and gray angelfish, and the butterflyfish, remain in pairs for extended periods and will always mate together. At completely the opposite extreme, some species gather together in vast "spawning aggregations".

Schooling species, such as blue tang, often have favored spawning sites within or close to their feeding grounds. These sites are typically in slightly deeper water, and usually near an outcrop or area where there is good mixing of the water. At one site in Puerto Rico a group of about 20 000 ocean surgeons was observed to gather and spawn every afternoon in the winter months.

Even more extraordinary is the gathering of the large snappers and groupers. Taking their cues from the water temperature and the state of the moon and tides, all of the large groupers, the smaller red hind, and the mutton, dog, and cubera snappers, set off on great journeys to spawning sites. Nassau grouper in Belize have been documented to travel for 240 kilometers to such a site. When they reach their destinations these fish form dense schools, like heavy thunder clouds. Even 500 or 1 000 large snapper in one place can be quite awesome, but it is estimated that before fishing pressure took its toll there may have been aggregations of Nassau grouper numbering up to 100 000 individuals. Spawning fish break off from these swirling masses in smaller groups, with 10 or 20 individuals shooting out, often towards the surface, and releasing a cloud of gametes into the open water at the peak of their ascent, before swimming back down into the maelstrom.

Changing sex

Many fish on the coral reef, including most of the groupers, parrotfish, and wrasses, have the ability to change sex at some point in their lives. They are born as either males or females, and it is typically the females which undergo such a secondary sexual change. This is often accompanied by an alteration in color and size, and in the wrasses and parrotfish such changes are very dramatic indeed.

Amongst these sex-changing species, society is strictly hierarchical. Dominant males typically defend a wide territory (or, in the case of the bluehead wrasse, they may just defend a small spawning site). Other dominant males are chased off, but the remainder of the population, which includes both females and young males (who have the same colors as females) remain. The dominant male seems to have exclusive spawning rights over his female harem, but studies have shown that the diminutive bachelor males often sneak in and mate with the females when the dominant male is not looking. Should a dominant male be removed from his territory, another fish from the harem will quickly take its place. Whether a natural male or a female, it will begin to behave like a dominant male, and within just a few days will undergo a complete transformation – sex, color, and size.

The pretty hamlets have taken sex changing in a slightly different direction. These fish maintain a stock of both eggs and sperm in the same individual. In their rather elaborate spawning rituals two hamlets will chase and interact for up to an hour, then one assumes a dominant, female role and they swim up, briefly intertwine, and spawn. A few minutes later this brief spawning ritual is repeated, but this time they reverse roles. There is actually a good explanation for this strange behavior – eggs are much more costly to produce than sperm, so by taking it in turn to release small amounts, each fish is ensuring that the other invests the same amount in the process.

Parental care

Many creatures on the reef care for their eggs, giving them a powerful head start over the eggs which are more extravagantly released into the open sea.

Most crustaceans have fairly complex routines both for getting together and for sexual reproduction. Spiny lobsters signal to one another with their antennae to show their interest in mating. The male then transfers a sticky mass, full of sperm, on to the female's sternum (breast area). The female may not release any eggs for days or months after this event, but when she does she scratches the sperm package and releases the sperm. She then holds the eggs on her underside close to her tail (large females may hold more than 500 000 eggs), for about three weeks. When they hatch she moves to a prominent position, swept by currents, to release the larvae into the plankton.

In quite a few fish species it is the males which care for the eggs and even, in the case of seahorses (see page 57) and toadfish, for the young. Damselfish males prepare a nest area, clearing it of debris and even chasing off such creatures as long-spined sea urchins (by nipping the tips off their spines). They entice females with a variety of maneuvers and dances, and the eggs are then laid on the ground. They take between three and seven days to hatch, and during this time the males waft them with water and pick off any dirt or debris. More importantly, they will viciously attack any other fish which approach to try to feed on the eggs. In some species of cardinalfish, as well as the jawfish, the males look after the eggs in an even more impressive manner – holding them in a great bundle in their mouths until they hatch.

Moving between reefs – life in the plankton

If there is one unifying factor in the life cycle of almost every animal on the reef, it is that the young spend part of their lives in the plankton. From sponges, to corals, to groupers, almost every reef creature releases eggs or young larvae into a free-swimming, or drifting, phase. These larvae often look utterly different from their parents, and, in the case of more sedentary creatures, they can be much more animated. Tunicates, or sea squirts, are considered to be close relatives of the vertebrates (animals with backbones, such as fish, and humans), not because of the appearance of the adults, which are sessile filter feeders, but because of their larvae, which resemble small tadpoles and have a notochord, similar to a spinal chord. Many fish larvae are active swimmers and predators, feeding on other members of the plankton.

This phase in the life history is critically important for coral reef survival – for most species it is the traveling phase of life. Once they settle, corals and sponges cannot move, and even active swimmers such as fish may never leave a home territory. But in the plankton, often for days or weeks, the larvae may be swept many miles, settling out on new reefs, and ensuring that there is a constant flow of new life from reef to reef.

Symbioses

Throughout the natural world, interactions between one species and another are common. The simplest such interactions are between predators

Thick clots of eggs rise up out of a spawning barrel sponge. All others on the reef will release their spawn at exactly the same time.

Gray snapper migrate to particular reefs over distances of tens or hundreds of kilometers in order to spawn.

and prey, but there are many other interactions and biologists use the term symbiosis (which means living together) to describe them. Quite often people use the term symbiosis to mean only the positive interactions where both species benefit, but it can mean any relationship, including some fairly nasty forms of parasitism.

Mutualism
This is the term more correctly applied to symbioses where both partners benefit. The invisible relationship between corals and their zooxanthellae is an excellent example, but there are many others. Among the easiest to observe

A group of cleaner gobies leave their station to attend to a yellowmouth grouper.

are the cleaning services provided by various reef creatures.

Most fish on the reef are prone to small parasites which attach to their skin. These parasites represent a potential food source for many smaller creatures and through evolutionary time certain species have developed skills as cleaners. The cleaners are represented by a host of smaller fish – notably cleaner gobies, but also many part-time cleaners, including bluehead wrasses and juvenile Spanish hogfish, porkfish, and French angelfish. Another gang of cleaners is found among the shrimps – notably the banded coral shrimp and the Pederson cleaner shrimp.

The cleaners often stay in fixed locations known as cleaner stations. These are visited by fish from across the reef, many of which take up particular postures or change color to signal their desire for attention. Barracuda will halt by a station, mouth wide open and body darkened with broad bars. Some jacks and parrotfish will rest with their bodies held vertically in the water, head up.

In addition to picking at parasites, the cleaners also clean up bits of dead skin, and will often swim right into the mouths to inspect the teeth of larger fish, and out through their gills, cleaning as they go. Injured fish have been seen to spend a lot of extra time at the cleaners, which presumably helps to prevent the wounds from getting infected. It is estimated that about half of all reef fish visit their services every day.

Commensalism
In quite a few relationships only one partner benefits, although the other does not really suffer. Hiding or seeking protection is quite common. Look closely into the protective spines of long-spined sea urchins and there are quite often small fish taking refuge. Juvenile jacks in the open ocean sometimes take refuge among the tentacles of large jellyfish. Stranger still is the conchfish, a tiny cardinalfish which takes shelter under queen conch during the day, but ventures out to feed after dark. But perhaps strangest of all is the slender and semi-transparent pearlfish – during the day this fish forces its way, tail first, into the anus of the sea cucumber, taking refuge there until nightfall when it emerges to feed on small crustaceans in the nearby water.

Another form of commensalism is that of

hitching rides. Larger fish are often seen to be accompanied by the elongate form of remoras. In these fish, the front fin has developed into a ridged plate which is actually a sucking device, enabling them to latch on tight to other fish. Remoras go wherever their hosts go. They are not always firmly attached, often just riding on the pressure wave of water which runs alongside any fish as it swims through the sea. They are often associated with sharks, but may be seen on turtles, groupers, snappers, even large parrotfish. They feed on plankton, small crustaceans, and on scraps dropped by their hosts, and they occasionally provide a cleaning service for their hosts. This is an uneasy relationship however, and remoras are quite often found in the stomach contents of sharks.

Remora – slender fish with a specialized sucking plate – are often seen on large fish such as sharks, getting a free ride.

Parasites

In some cases the relationship between two species is very one sided – one benefits and the other loses. This is parasitism. Many parasites are hard to see, but the wealth of cleaning stations across a coral reef bears testimony to their existence. The targets of many cleaners are tiny crustaceans and worms which attach to fish to feed on their blood, mucus, or body tissue. Internal parasites remain less well known, but, just as on land, worms and other creatures regularly infect the bodies of fish and other animals on the reef.

The boundaries between commensalism and parasitism are often a little blurred. When barnacles are observed on animals such as whales and turtles they appear to be doing little harm, but thick incrustations would clearly affect the streamlining of these species as they swim through the water. Another group, actually very common on the reef, are the isopods often observed on the flanks or faces of many reef fish. These large crustaceans appear to be horrific encumbrances, and they must certainly be awkward attachments, but they do not suck blood or otherwise feed on their hosts, and some even act as cleaners.

Taking it in

Life on the reef is always on the move, constantly changing. Human visitors, drifting over this marvelous world, can barely take in a fraction of what is going on. To begin to comprehend

A mated pair of isopods attached to a coney. Once settled they lose the ability to swim and remain on their host for life.

the reef, look for patterns, try to understand behavior, and learn the names of some of the creatures.

Look at the reef itself and consider its size, its shape, its age. Look at the living surface and try to identify some of the corals which built the reef. Look also at the numerous fish and invertebrates jostling for attention. The busy activities around a cleaner station; the roving gangs of feeding surgeonfish; the male wrasses defending their territories. Look for feeding animals, and see how far evolution has carried life on the reef – the wafting tentacles of fanworms, the powerful beaks of parrotfish and the needle-like teeth of barracuda. Armed with a little understanding, the fascination of the reef can grow with every visit.

The Wider Caribbean

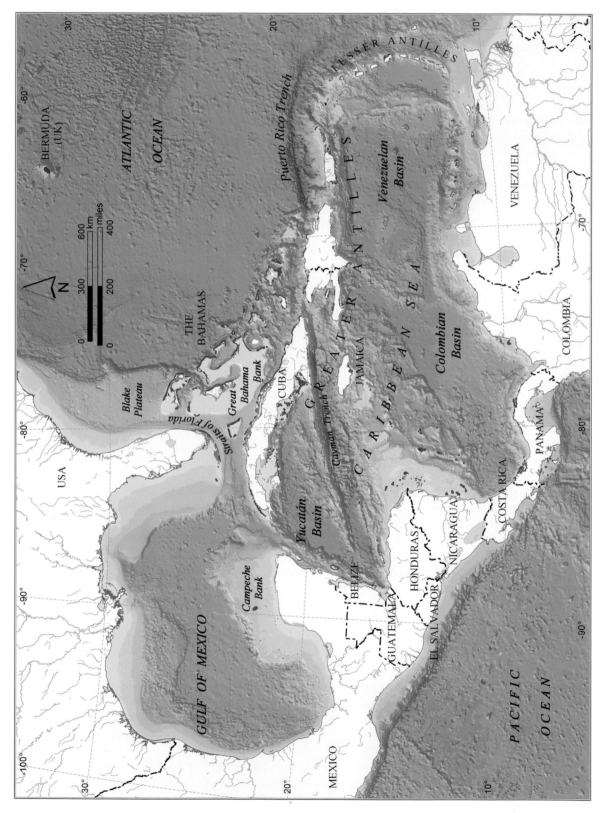

THE CARIBBEAN REGION

The Caribbean Sea is a large sea, closed off to the west and to the south by the Americas, and partly screened to the north and to the east by the island chains of the Greater and Lesser Antilles. Coral reefs are widespread throughout this sea: tight up against many island shores; in wide barrier reefs; and scattered like jewels across the wider continental shelves.

To the northwest the Gulf of Mexico forms another, distinct, oval basin of water surrounded by the southern coast of the USA and the eastern coast of Mexico. Although a warm sea with extensive shallow coastlines, there are only a few, relatively small, coral reefs.

Directly to the north of the Caribbean Sea lie the westernmost waters of the Atlantic. These are mostly outside of the tropics, but corals thrive here, supported by warm currents flowing from both the southeast and the west. This area of the western Atlantic Ocean includes the vast maze of shallow banks and coral reefs which make up the Bahamas archipelago. Here too is the Florida peninsula, edged by a great sweep of coral reefs stretching from the Straits of Florida into the Gulf of Mexico. Way off to the north, and extremely isolated, lies the island of Bermuda, home of the world's most northerly coral reefs.

Taken together, these three areas are sometimes called the Wider Caribbean. Their marine communities, and particularly their coral reefs, are intimately connected, and furthermore there are no other communities quite like them anywhere else the world. In this book we use the terms Caribbean or Wider Caribbean to mean this entire region, and reserve the term Caribbean Sea for the single central basin.

Old seas, young seas, and volcanoes

The origins of the Caribbean Sea can be traced back to a period from about 90 to 75 million years ago, when a large, oval-shaped fragment of the Earth's crust broke apart from its surroundings. The Earth's outer crust is made up of quite a number of these separate tectonic plates and their tiny, slow movements, taking place over enormous time scales, have shaped much of the surface of the planet. The newly formed Caribbean plate began to move slowly eastwards, pushing into the Atlantic and forcing up a great ridge of rock. In places this ridge rose clear of the ocean, forming islands and volcanoes and building a barrier that became the eastern edge of the Caribbean Sea – the Lesser Antilles.

Further plate movements began to shape the other boundaries of this sea. Mountains were thrown up, like a series of folds and twisting ridges, to the north and south as the Caribbean plate slid past the North American and South American plates, forming the Greater Antilles and the coastal ranges of South America. On the western side, the Caribbean plate was itself being pushed by the movements of the Cocos and Pacific plates. This threw up a great snaking ridge, which formed a permanent separation between the Caribbean Sea and the Pacific Ocean – Central America.

These movements are continuing today. Pushing eastwards at 1-2 centimeters per year, the Caribbean plate is continuing to crumple the Earth's crust, creating cracks and fissures, and allowing molten magma to reach the surface in a series of active volcanoes.

These slow-motion movements have also driven profound changes in the life found in these waters. All through geological history, reef life had moved freely through the oceans, but some 3 million years ago, when Central America rose up and connected the two continents, a permanent divide was placed between the Pacific and the Atlantic Oceans. The coral reefs of the Wider Caribbean became isolated from the rest of the coral reef world.

There followed periods of harsh climatic change. Great ice ages brought cataclysm to the coral reefs, wiping out nearly 80 percent of the Caribbean corals. But later, as conditions improved again, the remaining corals recovered and rebuilt the reefs. New species evolved to fill the gaps left by others. Separated from other coral reefs, however, these species were unique to the Caribbean, and this is why, even today, almost every species on a Caribbean coral reef is found nowhere else in the world.

Currents

Winds and currents exert a major influence on life in the Caribbean. Two great currents sweep towards this region from the Atlantic – the North Equatorial Current and the North Brazil or Guyana Current. The latter sweeps along the coastal regions of South America, picking up

The Grenadines in the Lesser Antilles typify the Caribbean coastlines, with warm clear waters, rich in marine life.

waters from the Orinoco River before reaching Trinidad and Tobago, and the southern islands of the Lesser Antilles, while the North Equatorial Current comes in more directly to the Lesser Antilles. These currents do two things. Part of their flow runs between the islands and into the Caribbean Sea, forming a generally westward flow across this sea known as the Caribbean Current. The remainder of their flow is deflected to the northwest, tracing a line past the eastern Greater Antilles and up past (and through) the Bahamas – the Antilles Current.

Over much of the Caribbean Sea the Caribbean Current is more of a gentle drift than a powerful force, and, like many currents, it sets up swirls and eddies. Thus, for example, back-currents sweep south and east along the coasts from Nicaragua to Colombia, and in a smaller loop around Belize and western Honduras. Eventually, however, these waters make their way towards the northwest corner of the Caribbean Sea where they are squeezed into the Yucatán Channel between Mexico and Cuba. Here there is a near continuous, and quite powerful flow, carrying Caribbean waters into the Gulf of Mexico. At this point the current is renamed the Loop Current, so called because it flows in a wide clockwise loop around the Gulf of Mexico, occasionally shedding off small eddies westwards towards the coast of Mexico. Finally, this water is swept towards another narrow passage, the Straits of Florida running between Florida and the Bahamas, and becomes the Florida Current. This surges out into the Atlantic, where it joins forces with the Antilles Current once again. These combined currents stream across the Atlantic as the Gulf Stream, and its warm waters bring a benign climate to the tiny island of Bermuda, enabling corals to grow further north here than anywhere else in the world.

Winds and storms

For the early travelers who crossed the Caribbean in sailing ships, among the best known features of the region were the northeast trade winds. These blow almost constantly throughout the year across most of the region. Trade winds are characteristic of many of the world's oceans and blow from areas of high pressure, which form just outside the tropics, towards the Equator. Their movement is always deflected

Hurricane Luis hits the coast of St Martin in 1995. Hurricanes rip through parts of the Caribbean every year, wreaking havoc on coastlines as well as underwater.

westwards by the rotation of the Earth (the Coriolis effect), so that, in the Caribbean, the trade winds always blow from northeast to southwest, bringing constant, sometimes quite strong, breezes to most areas. The same winds drive up waves, tending to make the eastern or northeastern shores more wave-swept, and greatly influencing the structure and development of coral reefs. On these windward coasts the reefs are often deeply incised, and corals are unable to grow in the shallowest water, whereas the leeward coasts are almost permanently protected and corals can build fine structures even close to the sea surface.

During the warmest periods of the year, especially when sea surface temperatures are high, tropical depressions may form. These are areas of low pressure with associated heavy winds and rains. In certain circumstances such depressions become highly concentrated into tropical cyclones, formed around a center of intense low pressure. Such storms have been given the name hurricane in the Caribbean, named after the Carib god, Huracan. Viewed from space they are circular in shape – great vortices formed around a central eye. The lowest air pressures ever

measured at sea level are found in these eyes, and powerful winds (a hurricane, by definition, must have winds of 117 kilometers per hour or more) race around their perimeter. They whip the sea up to a frenzy and the warm waters rapidly evaporate. The water vapor condenses out in great towering masses of cumulonimbus clouds, and falls in intense rain.

Typically, anywhere between 5 and 25 such hurricanes may form in any year. Physically they are intense, but not broad – averaging perhaps 400 kilometers across. They move over the region, sometimes strengthening, sometimes abating, cutting a swathe of destruction. On land, hurricanes may flatten trees and houses, while the intense rains lead to widespread flooding. On the coast, the low pressures cause dramatic rises in sea level, while the waves are amplified to heights of several meters, pounding the shores and the coral reefs. Such storms, in just a few hours, have been known to both build and destroy small coral islands. They may raze coral reefs to the sea bed. Hurricane Hattie swept over Belize in 1969 with winds gusting to 320 kilometers per hour, and the reefs in the central path of the storm took over 20 years to recover.

1.2 LIVING WITH REEFS

PEOPLE AND REEFS

The time of arrival of the first humans on the coasts of the Caribbean is unknown, but there is evidence of agriculture right across Central America by 7000 BC. Over time there was a great diversification of cultures and peoples. Many became intimately involved with the sea, traveling by boat and living off the abundant resources of the coasts and shallow waters.

The first peoples probably crossed over to Cuba and Hispaniola from around 5000-6000 BC, with a second wave traveling between the islands from about 1000 BC. Descendants of the latter formed the Ciboney people, who were still present in parts of Cuba when Columbus arrived in the 15th century. The majority of the island populations, however, descended from two later waves of immigration. First were the Arawak (also known as Taino or Lucayan) people, who

The Kuna people still have close ties to the coral reefs, living on coral islands and traveling by canoe to the mainland.

originated in Venezuela and spread up through all of the islands from about 300 BC. They were followed by the Carib peoples who moved through the Lesser Antilles from about AD 1000, displacing the Arawak as they went.

A host of different peoples lived along the mainland coast, and in contrast to the island inhabitants, many are still present today, largely in Central America. In Nicaragua, Honduras, and later in the Yucatán Peninsula (Mexico), the Mayans were among the most advanced civilizations of the Americas. Their origins can be traced back over 4 000 years, and from the 3rd century they developed towns and cities, hieroglyphic writing, considerable mathematical skills, and a calendar which was more advanced than that being used in the "Old World".

Although they had a few domesticated animals, many of these first peoples relied on the sea to provide them with animal protein. They occasionally took manatees, turtles, and monk seals, but more regularly ate conch and fish, which they caught with spears, nets, and baskets, as well as larger traps built in shallow water or across creek mouths. For the Arawak, one favored way of cooking was to stew food in a "pepper pot", that would be kept simmering – sometimes for days or weeks – with manioc, chilli peppers, vegetables. Such a means of cooking remains widespread in the region today. There is also a belief that the word barbecue may have an Arawak origin, and early Spanish travelers observed the Arawaks erecting wooden frames over fires in order to dry fish and meat.

Columbus first arrived in the Bahamas in October 1492 and was immediately entranced by this "New World". Over this and four other journeys he traveled extensively. His early descriptions, noted by the Dominican friar Bartolomé de las Casas, were full of praise for the land, the people, and even the marine resources. "Here the fishes are so unlike ours that it is amazing; there are some like dories, of the brightest colours in the world – blue, yellow, red, multicoloured, coloured in a thousand ways; and the colours are so bright that anyone would marvel and take a great delight at seeing them. Also, there are whales." (Cristóbal Colón, *El Libro de la Primera Navegación*).

Unfortunately the arrival of Europeans also brought one of the most dramatic changes ever experienced in human society. Within 50 years most of the original inhabitants of the Caribbean islands were gone. Many were taken as slaves and died in mines on the large islands and mainland America. Others died in battles with the arriving Europeans, but one of the most devastating factors in their demise was a susceptibility to European diseases – malaria, smallpox, tuberculosis, and typhoid swept through their populations and few survived.

European settlement proceeded apace, although with the loss of the original peoples there was undoubtedly a phase, lasting a hundred years or more, when human populations were lower than they had been for millennia. In a strange twist of fate this may have been a boom period for marine life – with very little fishing or hunting it is likely that populations of turtles, fish, monk seals, and conch would have increased dramatically across the region.

The development of colonial agriculture brought more changes, including the arrival of vast numbers of slaves from Africa. Forest clearance led to erosion, and the appearance of sediments in the coastal waters. Feeding the people in these new colonies was a challenge, and there were quickly signs of overfishing. Marine turtles were "mined" to virtual extinction in the Cayman Islands by the end of the 18th century, and the monk seal became completely extinct at around the same time. In a few places, such as Jamaica, there was a need to import fish from the early 19th century, as the reefs were unable to produce enough to feed the large local population.

Today, human populations are densely packed along most of the Caribbean coastline, and there is an immense diversity of cultures and peoples. In some areas local populations, whether of Amerindian, European, or African descent, live in close association with nature – fishing and farming. In other areas the changes have been radical – industrialization has brought mechanized agriculture, larger and more wide-ranging fishing vessels, and of course the development of industry, as well as large human settlements. Technology has also brought with it rapid international travel, and the Caribbean has become one of the world's major tourist destinations. The changes have been immense, and humans are now altering the environment faster than in any previous generation. At the same time, we are also learning just how much we depend on the natural environment, and are beginning to take measures to control the changes.

Fishing

Our knowledge of early fishing activities over wide areas of the Caribbean is dependent on a few early writings by European travelers, and on archeological evidence. Ancient refuse heaps tell an important tale – they are littered with conch shells, and with bones from parrotfish, grouper, and snapper. The people traveled easily by boat, and relied heavily on the coastal waters for food.

Today, fishing remains critically important. For some, the sea is still the main source of protein, and fishers go out regularly to catch food for their families, or for sale in local markets.

A MEDICAL TREASURE CHEST

Hidden amongst the still largely unexplored diversity of the coral reef may be many cures for human illnesses. As in all complex ecosystems, certain creatures on the reef, such as sponges, have developed complex chemical compounds which they use as toxins to deter attacks by other species.

Pharmaceutical companies have begun to investigate these compounds and there have already been some promising findings. Pseudopterosin, an anti-inflammatory compound, has been extracted from a Caribbean sea whip, and these corals are now being commercially harvested in the Bahamas. Discodermalide, with immunosuppressant and cytotoxic properties, was discovered in a deepwater sponge from the Bahamas. Ascidians, or sea squirts, in the Caribbean have yielded didemnin B, with immunosuppresive, anti-viral and anti-cancer properties, and ecteinascidin 743, with powerful anti-cancer properties. Similar anti-cancer properties have been found in a very simple blue-green algae, first collected from beaches in Curaçao.

Traps, lines, and nets are sometimes used from shore, but more usually from boats. Such fishing can be sustainable, but as coastal populations grow a point may be reached where there are simply too many fishers, outstripping the natural productivity of the reef.

In other places, fisheries have become heavily commercial, catching fish for high-value and export markets, and these are often even more difficult to sustain. Such markets tend to be fussy, demanding large numbers of just a few species, whereas coral reefs tend to have lower numbers of a wide range of species. The "target species", such as lobster, conch, snapper, and grouper, can be quickly overfished.

Recreational fishing is also widespread, not only in the wealthier Caribbean nations, but also among international tourists traveling across the region. These fishers target a host of different species, including game fish such as barracuda, jacks, and snapper, as well as marlin, sailfish, and tuna in the deeper offshore waters. The calm waters of the shallow lagoons are scoured by fly fishers eager to catch bonefish, permit, and tarpon.

Recreation

Tourism is regarded by many as the world's largest industry and for many decades it has been one of the fastest growing. In 1950 there were an estimated 25 million international tourist arrivals worldwide, but the World Tourism Organization estimates that this figure will be 40 times this, or 1.018 billion, by 2010. Such figures do not include domestic tourists, which in places such as Florida are numerous, and are growing fast in almost every country.

The Caribbean is one of the greatest centers for world tourism, and the vast majority of tourists come to visit the coast. Only a small proportion come specifically to see coral reefs, but snorkeling and especially scuba diving have become two of the fastest growing sectors in this industry. Meanwhile, even those who simply sit on beaches or swim in the coastal waters benefit from the reefs offshore, which provide the sand they sit on, and buffer the incoming waves.

In a few places coral reef tourism is the mainstay of the economy. This is particularly the case around some small islands – the Cayman Islands, the Netherlands Antilles, Los Roques (Venezuela), Cozumel (Mexico), to name a few. Without the coral reefs these economies would quickly collapse. In other places entirely new economies are springing up as dive tourists "discover" new locations. Fishers have often put down their nets in favor of work in hotels, dive centers, and marine parks.

Building coastlines

Although they may not know it, tourists lying on the white sand beaches of the Caribbean are lying on tiny fragments of the coral reef. Some of these fragments have been broken up by waves, but others have actually been turned to powder in the digestive systems of fish. Parrotfish and others regularly scrape and chomp at dead corals or at calcareous algae, making use of the living matter, and ejecting the rest in small clouds of sand. In fact the constant production of sand is an important factor in maintaining the healthy beaches appreciated by so many.

The reefs also play a more dramatic role. Fringing and barrier reefs provide important buffers to ocean waves, leaving tranquil waters along the coastal beaches. During the heaviest of storms these same barriers become critical. Suffering heavy pounding by the surf – and sometimes sustaining considerable damage – they prevent the same waves from reaching the shore. Of course not all coastlines receive this protection, and every year storm surges sweep away many buildings and even take human lives. Without the protection of offshore coral reefs, however, such damage would be far greater.

Diving and snorkeling are among the fastest growing sectors in the tourist economy across the Caribbean.

Understanding values

In a brave, but somewhat controversial study of the world's natural resources, it has been estimated that the world's coral reefs are worth some US$375 billion every year to human society. Such a vast figure stands a little lost on its own, and in reality it is far more useful to address the value of coral reefs at more local levels. But before doing so it is important to think about the question, and why we are asking it.

Coral reefs are under threat from a great range of human activities. At the same time, it is argued, they are of great value to humans. Almost every major decision affecting our human environment is influenced by arguments of economics, but few environmentalists have taken this into account. They have argued that places are beautiful, important, and diverse, but have rarely converted these into the economic statistics which could influence policy and development, and might greatly help to protect coral reefs.

For a single industry, such as fishing, it can be relatively simple to identify such values (for example the Bahamas lobster fishery was calculated to be worth US$60 million in 1997). It becomes more challenging, however, to calculate combined values for a wider range of uses such as tourism, recreation, subsistence, and industrial fisheries.

In Florida, a recent survey estimated that the coral reefs generate US$7.6 billion per year from combined recreational uses such as diving, snorkeling, and fishing. This works out at US$5 million for every square kilometer of reef in the state every year. Such a figure is influenced by the vast numbers of tourists, combined with the high costs of living in the USA. At the same time it should be realized that this figure does not include many of the hidden values of these reefs such as the creation of beaches and protection from storms.

Only a few studies are really trying to consider what they call total ecosystem valuation. One, in the Portland Bight Protected Area of Jamaica, considered the costs and benefits of protection for tourism, fisheries, forestry, biodiversity, and other sectors. They calculated a net value of US$10-14 million per year by 2025, but that this value would be only US$0.5 million if the site was left unprotected.

Economic arguments may go a long way in

Since the advent of scuba diving, scientists have been able to study coral reefs in ever greater detail, and they are providing fascinating insights into subjects from ecology to behavior, to evolution, to human health and welfare.

halting the damage being done to coral reefs, but such arguments only give a partial picture. Reefs provide food and employment, often in areas where there are few other sources, and often to poor people who have nowhere else to turn. In strictly economic terms, such people and such resources rarely rank highly, but human life and health has a value that is not easily transferred into dollars.

There are also powerful arguments about the intrinsic values of coral reefs, or of nature more generally. Coral reefs are places which generate deep and powerful emotions. Their sheer physical beauty cannot be captured in terms of dollars or jobs provided. They are places of solace or of inspiration, places of awe and excitement. These are weak arguments in the face of development. They rarely stand up in court, and many economists still fail to take them into account. But as society develops, and as the natural world becomes challenged on all sides, the last remaining wildernesses are becoming critical, many would say, to the health of human society.

THREATS TO REEFS

This book adopts a positive outlook on the state of the Caribbean's coral reefs. But it must be remembered that many coral reefs right across this region are a shadow of their former selves. Some have died completely, and lie buried in layers of silt, crumbled by storms, or overgrown by algae. Others are still hanging on, but depleted of fish. Corals have died even on the most remote and unpolluted reefs, and almost everywhere there are far fewer corals today than there were even 10-20 years ago.

Here we examine the causes of these problems. Some are past, some are still occurring, and some are perhaps best considered as potential threats, happening already in a few areas, but hanging over others as yet unspoiled.

Diseases

In 1983 an unknown disease which killed the long-spined sea urchin was first observed in the waters around Panama. The mystery pathogen appeared from nowhere, but within a year had spread through the entire Caribbean and killed over 90 percent of these sea urchins everywhere. There have been suggestions that it could have come through the Panama Canal, but in reality we have no idea where it came from.

In quite a few places, but especially where there were problems of overfishing, the long-spined sea urchin was actually the most important grazer of algae on the reefs. As the urchins disappeared, the algae began to flourish, sometimes in direct competition with the corals.

Black band disease on a common sea fan.

Diseases on corals were first reported in the 1970s, but it was not until the 1980s and 1990s that some of them began to have a dramatic impact. Until this time two corals in particular – staghorn and elkhorn – were the most abundant and important shallow water corals in the Caribbean, and it was their skeletons, more than any others, which were found in the thick limestone deposits of older reefs. These were the major reef builders. But today most reefs have lost 80-100 percent of their staghorn and elkhorn coral through a disease known as white band disease. A host of other coral diseases have been reported, often named after their appearance on the corals (black band, white plague, yellow band, dark spots, and rapid wasting disease).

Remarkably little is known about these diseases. In some cases we do not even know what type of organism may be causing them. We do not know where they first appeared, nor what the initial cause may have been. One theory is that one or more of them may have come from fungal spores blown into the Caribbean from the Sahara Desert. Although this seems far-fetched, the loss of forest and the increasing desertification of Northern Africa has led, on a number of occasions, to large amounts of red African dust being transported across the Atlantic in the upper atmosphere before falling out over the Caribbean. Algal spores similar to those which have caused some of the coral diseases have been found in this dust.

Coral diseases appear even in the most remote and healthy reef communities, but they are probably more common, and certainly more devastating, in areas where the corals are already stressed by pollution or sedimentation.

Sediments

Corals thrive in clear water. If the water becomes thick with mud or silt two things happen. Firstly the sediments block out light, so the algae within the corals cannot photosynthesize so well. Secondly, the sediments sink down, falling on the corals like rain. The corals are able to use their tentacles to remove these sediments – sloughing them off with a layer of mucus, but they cannot keep doing this. It uses up energy and precious resources. In a few places corals have adapted and grow relatively well in areas of some sediments, but for the most part any dramatic increase in sediments will kill corals.

Unfortunately, all over the Caribbean there are increasing levels of sediments in coastal waters. The clearance of forests and the movement of agriculture on to steep slopes have been major causes of this, as both lead to soil erosion, with the soil particles being carried by rivers out into the sea. Chopping down mangroves adds to the problems, as mangrove forests are places where many sediments settle out in the still water.

Pollution

The most widespread form of pollution found on coral reefs comes, not from toxic wastes, but from simple nutrients: organic matter, nitrates, and phosphates. Two of the major sources of these in the Caribbean are sewage and agriculture. While human waste and fertilizers provide a healthy food supply for some organisms, they do not suit corals. Instead, other organisms – starting with plankton and algae – move in and thrive. In sufficient quantities plankton forms a rich soup in the surface waters, preventing the light from reaching the sea floor. Even if the plankton does not bloom, the larger types of algae, or seaweeds, proliferate on the sea floor where their growth can out-compete the corals.

Toxic wastes also create problems, though on coral reefs their impacts tend to be quite localized. Sources include detergents, paints, herbicides and pesticides, industrial toxins and oils. These often come from factories and ports, from ships in transit, and from intensive agriculture, but marinas, golf courses, and hotels, often close to the coast, also add to the problems in many areas.

Yellow blotch disease on a star coral.

Overfishing

There is some evidence that even the early Caribbean peoples sometimes took too much from the oceans, causing localized collapses in what was available to eat. Nothing, however, has prepared the natural world for the onslaught of overfishing which has affected the world's oceans over the past 30-40 years. Nowhere has escaped, and the United Nations estimates that only about a quarter of the world's fish stocks are still in anything like a healthy state.

Broadly, the Caribbean fisheries can be divided into the capture of high-value fish mostly

Wide fields of staghorn corals across the Caribbean have been devastated by white band disease.

for export (and hotels, etc), and more local fisheries. The former includes the collection of lobster and conch (fisheries experts categorize these as "fish") as well as groupers, snappers, shark, tuna, and billfish (swordfish, sailfish, marlin, and their relatives). Almost everywhere these fish have been driven into rarity. Fishing for local consumption often targets just about every fish large enough to eat – Jamaican fishers catch more than 200 different species. Fishing in this way can be a little more sustainable, but as human populations grow, so demand quickly begins to outstrip the productivity of the oceans.

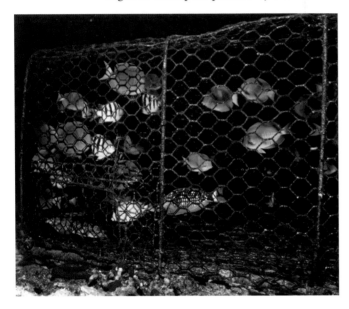

Fish traps are widely used throughout the Caribbean. When dropped on to coral they can be very destructive, and narrow meshes can lead to the capture of immature fish.

Unfortunately, although we know what to do about overfishing, it has proved incredibly difficult to persuade both politicians and fishers to bring about change. The result is that fish are reduced to very low numbers. Fishing continues, but there are not enough fish to reproduce and restock the reefs, which therefore remain quite unproductive. Well managed, the same reefs could supply more fish and would also be more attractive to tourists.

One of the saddest excesses of the fishing industry has been the capture of some of the largest fish when they gather in spawning aggregations. These rare gatherings of giant groupers and snappers are of fundamental importance

to the reproduction of these species, but fishers have treated such events as bonanzas. By so doing they have killed the "goose that lays the golden eggs", exterminating the populations of species such as the Nassau grouper in many parts of the Caribbean.

Direct destruction

For the most part, human actions have degraded rather than destroyed coral reefs, and total annihilation is rare. At small scales, however, physical destruction is common. Boat anchors and dragging chains are a continuous threat, particularly where boats are numerous. The sheer weight of boats off Florida means that there are dozens of direct collisions of boats with the reef, and outboard motors regularly churn and cut their way through seagrass beds.

Dredging and land reclamation have also taken out some reefs. Near the mouth of the Panama Canal huge areas of coral reef were literally dug up and then piled into other areas of shallow water to create new land. In a few places, especially along the coast of Florida, beach nourishment is another damaging activity. This is where sand is dredged up from offshore, or even imported from elsewhere, and is dumped on the beach, sometimes simply to enhance its appearance, and often to try to counter the effect of erosion. Such dumped sand is rapidly re-arranged by the currents and tides and can move offshore where it smothers the reef.

Coral bleaching

For reasons not yet fully understood, most reef corals appear to be very sensitive to changes in temperature. Living in the tropics, they have adapted and flourished in a world of very stable temperatures, varying little from day to night or season to season. High temperatures, even just 1 or 2°C above the normal summer maximum, drive a stress response in the corals known as coral bleaching.

When this happens, the stony corals become much paler or even white, caused by the loss of their symbiotic algae. These algae normally provide the corals' colors, and when they are lost the bright white skeleton of the coral begins to show through. Even when pale or white like this, the coral is not dead, but without its algae it is severely weakened and it may be

Sediments swept from the land rapidly cover reefs and kill corals.

more susceptible to disease. If conditions do not improve the coral will die.

Such bleaching is an entirely natural response to a seemingly natural stress, but what is causing grave concern among scientists is that the world's oceans are beginning to warm. Over the last century the temperatures across the world increased by about 0.6°C. Most of this occurred in the last 20-30 years, and the rates of change are now accelerating.

These are the first signs of global climate change and, while there are still disputes amongst scientists about the exact patterns of predicted change, the fact of climate change is no longer argued. Of course natural fluctuations in the environment will still create individual warmer or colder years, but underlying these fluctuations is a basic trend: the Earth is getting warmer.

Coral bleaching was a rare and isolated phenomenon until 1979, when numerous corals in the Pacific and the Caribbean showed the symptoms for the first time. Since then there have been observations of bleaching with increasing intensity every time there is a particularly hot year. In 1995 a mass bleaching event was observed on the Belize Barrier Reef –

numerous corals were affected, particularly in shallow water. A similar event in 1998 was compounded by the impacts of a hurricane, causing widespread loss of living corals.

Scientists are very concerned that corals will bleach more and more regularly as sea temperatures increase over the coming decades. The most gloomy outlooks predict widespread coral death, leading to total loss of most coral reefs within as little as 30-40 years. More optimistic outlooks suggest that corals will be more resilient – the survivors of early events are obviously better adapted to heat stress, and will repopulate the reefs with more hardy corals. If we can reduce some of the other pressures on reefs, and act rapidly to protect both the most sensitive and the most resilient areas, we may be able to help in this process. Even the most optimistic, however, predict that there will be devastating, and largely unpredictable, impacts over the coming years.

Lethal cocktails

In lining up the many threats facing coral reefs it is important to realize that none of these acts in isolation. Many reefs are already beset by overfishing and pollution, so that when a bleaching event or a hurricane comes along this may be just enough to tip the reef over the edge into a steep spiral of decline (see page 135).

At the same time, it is important to keep matters in perspective. Even degraded reefs have much to offer in terms of fishing and diving. And these same reefs can quickly be restored to their former glory if conditions improve, and if we offer them better protection.

Anchors and chains quickly pulverize the reef to a mass of rubble and shifting sand.

PROTECTING THE REEFS

Countless people rely on coral reefs for their livelihood and well-being but, despite their importance, the reefs are being damaged almost everywhere. These degraded reefs are failing to deliver the values – in terms of dollars, or jobs, or food supplies – which healthy reefs can offer in abundance.

One of the most important tasks that must be undertaken in order to improve coral reef protection is education. There is a need to show people, from politicians to fishers, that methods of protecting coral reefs exist, that they provide great benefits, and that they really do work. Alongside education, active management must be put into place, and this is often most effective if the people who use the reefs are directly involved in the management process.

Education

It is critical that we get the message of coral reef protection across. Coral reef ecology and management needs to be taught to school children. But fishers, planners, and politicians also need to be informed. Teachers, scientists, and environmental groups all have a role to play in educating, but so do the communities themselves. In a few cases, fishers who have direct experience of successful reef management have been taken to talk to fishers in other countries. The trust between people who do the same work is highly effective in convincing otherwise skeptical fishers that things can be done better.

Working together

A lot of people have a stake or an interest in coral reefs. Some know it, such as fishers and travel industry workers. Others do not feel so directly involved, but still enjoy visiting the reefs. Yet more are silent beneficiaries, eating the fish or enjoying the coastal protection, without realizing that they too are users of the coral reef.

Given responsibility, many of these – village groups, fishers, dive centers, environmental groups, or local governments – can manage their own coral reefs extremely well. Even more effective is when they can be brought together. Sometimes there are conflicts between different users, but if these conflicts can be worked through it is far more likely that a robust system of management will be set up, that will be obeyed and respected by all. In parts of the USA, recreational fishers have a huge mistrust of conservation, but in other places recreational fishers and conservationists are working together to build up new protected areas, offering some of the best fishing – and some of the best diving – in the Caribbean.

Fishing

Overfishing is widespread, but can be easily solved using a range of controls. In all areas, damaging fishing methods must be stopped. Such methods include the careless use of traps and nets which can smash corals. Even more damaging, though thankfully not so widespread, is the use of explosives or bleach to drive fish out of the reef.

Some management measures try to limit the size of the catch, the species caught, or the size of the individuals captured. For example, minimum mesh sizes on nets or traps can be set, allowing immature fish to escape. Closed seasons – periods when particular species may not be caught – help to protect spawning stocks. In some places the number of fishers may need to be reduced, or new fishers prevented from joining the fishery.

Bringing such measures into play can be very difficult, and it is usually necessary to work with the fishers themselves. In a few cases, government subsidies may be used to ease financial losses, or even to buy out fishers. One of the most encouraging tools for fisheries management, however, is the closure of certain areas as no-take zones (discussed below). These are now being used in a number of areas and have revolutionized fishing in many communities.

Land-based problems

Many of the threats facing coral reefs can be traced to land: from the sewage outfalls of nearby hotels to the remote farming practices high up in mountainous areas. Many of these require quite active involvement by governments or other bodies, controlling building, forest clearance, and waste disposal.

Innovative measures such as pollution taxes can encourage better wastewater treatment, while financial incentives, and indeed the sharing of benefits derived from tourism or other gains, can help to turn around the problems of sedimentation and pollution which could otherwise threaten coral reefs.

Protected areas and no-take zones

In some parts of the world, communities have long known that closing certain areas to fishing is an extremely valuable management tool. Such methods are still practiced in traditional societies in the Pacific, but more recently the modern world has begun to revive these approaches with the establishment of no-take zones.

Examples from across the Caribbean have shown that if all fishing can be halted on even a small part of the reef, the fish will thrive and will spill over into the surrounding areas. Over the first one or two years the fishers may loose a proportion of their catch (some schemes have compensated them for this loss), but after this, the protected stock in the no-take zone begins to replenish and restock the surrounding reef to much greater levels than before. Fishers start to "fish the line", lining their boats along the edge of the no-take zone, but even far outside these zones they find that catches are greatly increased. The science here is rock solid. Benefits continue to grow until about 30 percent of the reef is closed to fishing, simply because the large stocks, and indeed the large fish, in the closed areas are so abundant that they overflow into the surrounding fishing areas, while their larvae repopulate the reefs over even wider areas.

In some places these same no-take zones, packed with large fish and healthy coral, have become highly sought-after dive locations, and some fishers have given up fishing in favor of more lucrative or rewarding work in tourism. The tourists are often willing to pay for the privilege of diving in the reserved areas, and these fees

Well marked marine reserves, respected by local people and properly enforced, can rapidly restore fish populations.

have been used to help pay for the management of the sites.

Of course fisheries management is only one reason to set aside areas of the ocean. Many national parks and nature reserves have been established both for the enjoyment of people and for the protection of wildlife. Such protected areas were slow in coming to the marine environment, but from the 1980s onwards there has been a burgeoning of protection right across the Caribbean.

No-take zones form part of this network, but many of the other, often larger, protected areas offer differing levels of protection, dealing with certain types of fishing, boat anchoring, waste disposal, and so on. Unfortunately some of these protected areas are "paper parks" – they exist in law, but are not enforced. In others the regulations are too weak to provide much real protection. Design can be a problem too – few parks provide protection from threats such as pollution and sediments arising beyond their boundaries. But in amongst this network there are some real gems.

Many sites, such as the Soufriere Marine Management Area in St Lucia, are "zoned", with some areas declared as no-take, while others are open to recreation or to fishing. Some, such as the Land and Sea Parks in the Bahamas, have included large parts of the land as well as the ocean, offering the potential to prevent threats from inland sources. In these and a few other cases, the process of setting up protection has involved consultation with the full community, and these are actively involved in the ongoing management of the sites, as is the case with the Portland Bight Protected Area in Jamaica.

Integrating management

Ultimately the natural resources of any country must be managed as an integrated whole. The establishment of a banana plantation far inland cannot be allowed to destroy the livelihood of a fishing community 20 kilometers away. The development of a tourist facility cannot be allowed to threaten the last turtle beaches in a country, not only exterminating the turtles, but actually reducing the value of the diving industry.

Protected areas are not enough to achieve this, so many environmental groups are now asking for vastly more connected management. Governments need to trace or to predict all of the impacts, positive and negative, of any activity. Forestry, planning, and agriculture departments need to talk to fisheries, tourism, and environment departments. Such integrated coastal management also requires giving the people a voice – communities, fishers, and environmental groups. A few governments, such as that of Belize, have recognized this need and have established a coastal zone management program; however for most there is a long way to go.

International action

Many reefs run across international borders, and the shifting patterns of winds and currents build tight connections between the coastal resources of most Caribbean nations. With this connectivity, human actions in one country often have ramifications for its neighbors. The plundering of spawning aggregations of groupers, for example, will affect grouper survival over thousands of square kilometers. For some threats, such as climate change, dealing with these problems will require a truly global effort, but others are regional and there are growing efforts to get Caribbean nations to work together.

Some of these efforts are being coordinated through the United Nations, particularly through its Caribbean Environment Programme in Jamaica (see the list of environmental organizations on page 241). This center coordinates the implementation of the Convention for the Protection and Development of the Marine Environment of the Wider Caribbean Region. Also known as the Cartagena Convention, this is the major international environmental agreement for the Caribbean. It has been signed by most nations, and has important protocols encouraging countries to combat oil spills and land-based sources of pollution, and to establish protected areas.

Other collaboration is being developed in different parts of the region. A number of international projects have begun to focus attention on the Mesoamerican Reef region. In the Lesser Antilles different groups are trying to work together to manage fisheries in a more sustainable manner. Although many have a long way to go, these efforts may be the only way of helping to protect coral reefs and other marine habitats from the widespread threats that encompass large parts of the Caribbean.

SHIFTING BASELINE SYNDROME

One of the key tasks facing those who are trying to improve the marine environment is to raise people's hopes. All too often we forget how good things can be. The arrival of Columbus' ships in the Cayman Islands challenges the imagination – how his boats had to almost plow their way through densely packed turtles in the shallow waters. Many divers today thrill to catch the briefest glimpse of a fleeing shark on a reef – they have no idea that sharks were once regular company on almost every dive. Young fishers scoff as the old men tell stories of how big the fish used to be.

Scientists have devised a term for this disbelief in what once was: they call it the shifting baseline syndrome – any generation establishes a sense of what is normal based on what they have seen with their own eyes. One danger here is that it makes us rather unimaginative as we set our aims for conservation. We think "if only we can keep it how it is", or perhaps we aspire to seeing a few slightly bigger creatures and a few more corals.

We are an order of magnitude too low in our expectations, and thankfully there are still some places that can reinforce this message. It is still possible, on the Silver Banks for example, to sit in a boat surrounded by humpback whales. Divers can be followed about by enormous jewfish and continuously buzzed by sharks in southeast Cuba. There are beaches in Central America where over 10 000 turtles still come up to breed each year. The aggregations of spawning snappers off the coast of Belize are so large that schools of the world's largest fish, the whale shark, come in to scoop up the eggs in the plankton. This is the environment we should be aiming towards restoring everywhere.

WHAT YOU CAN DO

Although governments, managers, and industry have the power to bring about changes to help coral reefs, they need an incentive. Politicians will only do what they think the public wants, and businesses will only do what their customers demand. If we are to succeed in turning things around, it is critical that we change the attitudes of the public, starting with ourselves.

One of the most important things we can do for coral reefs is to get out and enjoy them, and to encourage others to do the same. They must not just be the realm of foreign visitors, but must be accessible to all.

For regular visitors to coral reefs, there can be few activities more rewarding that taking a newcomer to see a reef for the first time, sharing in their excitement and awe, and pointing out the different animals of the reef. Even non-swimmers can be taken snorkeling, using buoyancy aids to float over shallow reefs. Every visitor to a reef is transformed, and enthusiasts become advocates.

GET ACTIVE

√ **Join an environmental group.** (See page 241.) These are the key agents for change, lobbying leaders, and taking practical action. They are also great ways to meet people and to learn more about the environment.

√ **Take someone snorkeling.** The more people who enjoy coral reefs, the better chance we will have to protect them.

√ **Work underwater.** Quite a range of groups (see main text) are looking for volunteers to take part, or support science or management on the coral reefs.

√ **Write letters.** Letters to politicians, hotel managers, dive centers, or heads of industry can really make a difference. Also highly effective is to write to newspapers, or to dive or travel magazines. Make sure you copy any such letters to the "target" (politician, manager, tourist office).

√ **Clean up.** If you see trash on reefs or on beaches, take some home – remember turtles and many species of fish are killed by plastic.

√ **Live sustainably.** Think about what lies downstream.

• Use biodegradable products.
• Reduce waste – don't buy overpackaged goods, and dispose of all waste properly.
• Recycle, and buy recycled goods.
• Use less water. Even in water-rich countries, tap water is costly to produce and contains chemical additives. Its disposal adds to sewage treatment loads.

√ **Combat climate change.** One of the greatest threats to coral reefs, climate change is largely being driven by the inhabitants of developed countries whose lifestyles demand excessive amounts of energy.
• Turn off lights, monitors, air conditioning, fans, or heating if you are not in the room.
• Do not overheat or overcool your home. Dress for the climate.
• Avoid air conditioning. Wear less, open windows, and use fans. Many tropical buildings have been designed to enhance breezes.
• Drive less, and drive a fuel-efficient car.
• Use public transport, or walk, or cycle.
• Encourage sustainable energy generation – wind, water, sunlight, and even tides offer great opportunities for cheap energy at no cost to the planet.
• Buy local produce to reduce energy consumption from long-distance transport.

Of course with ever increasing numbers of people taking to the reefs, there is a danger of loving them to death. Divers and snorkelers cause great damage if they touch animals and break corals. Boat anchors can plow up large tracts of the sea floor. Many hotels fail to treat their sewage, and some have been built in former mangrove forests, or have built walls in the sea to protect their beaches. Fishing is fine if it is well managed: over most of the Caribbean it is not. The boxes presented here provide some guide-lines for halting negative impacts to the reefs, and for helping to restore their former glory.

Divers and snorkelers have a particular responsibility – when they visit the reefs their presence can be a source of harm, or a force for good, depending on how they behave. Many operators are now highly sensitive to marine conservation, but don't rely on others to keep you in line – follow the guidelines set out below.

Fish feeding is not fully addressed here because opinions are still a little divided on how

ZERO IMPACT DIVING AND SNORKELING

√ **Be comfortable.** Whether snorkeling or diving, make sure you are comfortable and that your equipment is properly adjusted. This will in-crease your enjoyment and prevent damage from clumsy movements in the water.

√ **Control your buoyancy.** Many beginner divers struggle to stay at one level in the water and move up and down constantly. Until you have good control of your buoyancy stay high above the reef.

X **Do not over-weight.** Many divers, especially early on, like to feel heavy, but by constantly sinking down you may crash into corals and other life on the sea floor.

X **Never touch anything.** Be particularly careful of corals and sponges – touching can damage their sensitive surfaces and lead to lesions, infections, or death. (Also, some reef animals sting or scratch.)

X **Don't chase fish.** Rapid movements in the water tend to scare off many fish. Some of the best opportunities for fish watching come from patience and gentle movements. Many fish are curious, and by holding still, or approaching at an oblique angle, you will encourage them to stay still or come to you.

X **Don't hassle other creatures.** The temp-tation to see what an animal does if you poke it or chase it can be strong, but the rewards are poor – most animals simply race off or hide. And they will be even more wary of the next diver.

X **Don't let your equipment drag.** Divers in particular often have long consoles that can dangle downwards and unwittingly smash into life on the sea floor.

X **Don't wear gloves.** It's a useful injunction to prevent you thinking you can touch things. Many dive centers operate a no-gloves rule.

√ **Stop and just take it all in.** Too many visitors rush about looking for the "big stuff" or the dramatic scenery. Everywhere you go on the reef there is an abundance of life. Look up close, or stop and just look around, 360°, to admire the view.

√ **Ask yourself questions.** Try to work out what's going on, what a fish eats, or why it behaves in a particular way. Learn a few names of different reef creatures. Armed with this knowledge you will find the reef even more fascinating.

√ **Find an area of sand or rubble** if you need to touch ground.

√ **Be especially careful taking photos.** There is a tendency to become so focused on what is in the viewfinder that you can knock into the reef itself – some photographers even use the reef for support or try to rearrange scenes for effect. Don't!

√ **Be willing to tell others** if they are doing something wrong. Or talk to your dive masters.

√ **Support a dive center or tour operator that is clearly promoting the environment.** Many are active proponents of conservation and by choosing to dive with them you will also learn more about reef life.

BOAT OWNERS

✗ **Don't anchor near reefs.** Anchors destroy corals and damage seagrass beds. Where available, use mooring buoys, and use a long leader line so that your boat doesn't lift the buoy out of the water.

✗ **Don't pump sewage** within 5 kilometers of a reef. In areas of heavy boat traffic don't release it into the sea at all. Use pump-out facilities in ports and marinas.

✗ **Don't pump bilge or gray water** (from showers and sinks) in shallow or enclosed waters; when possible use terrestrial pump-out facilities. Oil, detergents, and other chemicals are very damaging.

✓ **Travel slowly where people may be diving.**

✓ **Travel slowly where there may be manatees.** Collision with propellers is one of the major causes of death for these gentle giants.

✗ **Don't dispose of solid waste in the ocean.** Be wary of the claim that "the fish will eat it" as an excuse to dump organic matter into the sea. Coral reefs work perfectly well without a rain of human waste.

✓ **Use oil-absorbent sponges and rags** to soak up leakage or overflow in the bilge, or when filling tanks, and dispose of these on land. Oil kills!

✓ **Use biodegradable detergents,** especially when cleaning decks and exterior, and only do this far from reefs.

✓ **Avoid toxins.** Almost all paints, varnishes, thinners, strippers, bleaches, and the like are toxic. Try to find environmentally friendly alternatives, and even then only apply them far from reefs and other sensitive ecosystems.

FISHERS

✓ **Obey the rules.** These vary considerably from place to place, but are there for a reason. Check out closed areas or seasons, restrictions on gear or catch size. Some species are protected year round.

✓ **Be careful with fish traps.** Make sure they have biodegradable panels, so that they won't carry on ghost fishing if you lose them.

✗ **Don't use nets or fish traps over coral.** Apart from possibly losing or damaging your gear you will damage the reef.

✗ **Don't use spear and scuba.** This form of targeted hunting has decimated the great giants of the reefs, such as jewfish. There needs to be a basic stock, out of reach, in order to sustain life on the reef.

✗ **Don't kill what you don't need.** Many fish do not make good eating, and won't sell in markets. Put them back – most fish survive the trauma of being caught if they are quickly released.

✗ **Don't take undersized or gravid individuals.** Animals must have a chance to mature and breed if there are to be any for future fishing trips.

✗ **Don't fish spawning aggregations.** These are fishers' security for the future. Some countries have destroyed them, and with them the livelihoods of fishers.

✓ **Consider tagging.** As an alternative to killing large game fish, many operators encourage, or even insist on, simply photographing and weighing the catch, before releasing it again, tagged as a proof of capture.

✓ **Put back what you can't eat.** There's something great about catching for the pot, but be responsible. Take only what you really want – and in places where fishing is heavy perhaps you shouldn't even take that.

✓ **Take waste home.** And look out for other people's, as well as your own.

to deal with it. In a few countries, and many marine parks, fish feeding by divers and by snorkelers is banned. It disrupts the natural processes of life on the reef, and certain foods may even endanger the lives of fish. Despite this there are counter-arguments, which point out that fish feeding is very localized, and that it attracts astonishing numbers of fish, creating major tourist attractions. Allowing fish feeding at specified sites may promote interest in reefs, while forbidding it in other areas allows reef life to continue in a more natural state where divers and snorkelers can experience the normal ebb and flow of reef life.

Fishers and boat operators, like divers and snorkelers, are on the front line. Contrary to some views, fishing and conservation can work extremely well together. It is the fisher's aim to catch fish, and properly applied conservation measures almost invariably lead to a greater avail-ability of fish to catch. Of course obeying the rules is critical, but in the absence of rules (few countries have adequate controls to prevent over-fishing) there are many common-sense measures which will ensure that fishing today will not affect fishing in the future.

It is not only in the water, however, that we can make a difference. Choices of hotels, meals, even souvenirs, all have an impact on coral reefs. Again, the guidelines laid out here are largely based on common sense. There are growing efforts to help consumers choose environmen-tally sustainable products, notably through the setting up of certification schemes. These are intended to give independent verification of en-vironmental sustainability – hotels or suppliers seek certification of their products, and consu-mers, on seeing the certification, are reassured that what they are buying reaches certain standards. In the coral reef world three areas of certification are beginning to open up. Watch these closely, and encourage them.

In the aquarium trade, the Marine Aqua-rium Council is now operating a very useful scheme providing certification to suppliers whose products are being caught in a non-damaging and sustainable manner. With these provisos the aquarium trade may be a very good thing for coral reefs – bringing high incomes for a low-impact form of fishing and giving the fishers an incentive to protect their reefs. Efforts to provide certification for fish as food products – led by the

EATING OUT, BUYING SOUVENIRS

√ **Think.** If you are in a crowded resort location the chances are that demand outstrips supply. Lobster, conch, grouper, and snapper are among the first fish to disappear, and so unless you are sure there are still healthy numbers in a place it is best to avoid all of these.

√ **Ask** how well the fishing is managed. It may be hard to get a straight answer, but if in doubt avoid seafood.

√ **Avoid undersized fish.** Of course it is difficult to judge, but as you learn more about fish, you will get better at telling what size an immature fish will be. For lobster, a tail length of less than 14.5 centimeters or a carapace (main head and body segment) of less than 9.5 centimeters is too small.

✗ **Don't eat swordfish, shark, or shrimps.** They are being overfished in all the world's oceans. Shrimp (or prawn) fishing is doubly destructive –

where caught in the wild the trawlers tear up the sea floor and catch, and kill, vast numbers of fish which are simply discarded. Where shrimp are farmed, almost every farm is responsible for the destruction of mangrove forests and the pollution of surrounding waters.

√ **Walk away** from any restaurant that offers turtle meat.

√ **Buy local goods.** Paintings and handicrafts are a great source of employment and help to put money into the local economy.

√ **Watch out** for products that have come direct from the sea. Never buy corals, seashells, inflated porcupinefish, seahorses, or turtleshell products. For some of these, including turtle products and ALL corals, it is actually illegal to carry them across international boundaries. They are protected under international law.

Marine Stewardship Council – are also on the increase. Unfortunately these have not yet moved into certifying reef fisheries. In the meantime, however, there are several groups which advise more generally on what should or should not be eaten (see page 247).

Although a number of certification schemes have been established for the hotel trade, these schemes mostly lack credibility – many offer certificates simply based on promises, not on action or achievements. It is still worth encouraging certification, but be aware that the current systems for hotels are poor, and many of the most environmentally friendly hotels have no certification at all.

Getting involved

Quite apart from being sensitive divers, fishers, or travelers, many people are seeking to do more to protect the world's coral reefs by becoming more actively involved. One particularly valuable way is to become involved with local environmental groups or dive centers in such activities as trash clearance – on reefs or beaches – or helping education efforts with local schools.

Increasingly, divers and snorkelers are also being asked to make scientific observations of fish and other life on coral reefs. Probably the largest organization of its kind is Reef Check, which has national coordinators in more than 20 countries across the Caribbean and is playing a very important role both in educating people and in monitoring coral reef health.

Those with a little more time and some financial resources may choose to join any of the organizations which are training and then using divers to gather information on coral reefs in the more remote regions of the Caribbean. Divers are camping on the beach in Andros and surveying the waters of the new marine park; others are staying in research camps in Belize, Honduras, Jamaica, Nicaragua, the Virgin Islands, and many other places.

Volunteers, whether observing manatees or damselfish, find that the learning process is enormous, but their contributions are also invaluable. These organizations have furthered not only our understanding of reef science, but have helped governments in the establishment of protected areas, and provided important interaction with local communities, stimulating a growing interest in coral reefs across the region.

CHOOSING A HOTEL

Considerable damage has been done to the coral reefs of the Caribbean by the building and running of hotels. Seeking exclusive beach frontage, new hotels push ever deeper along unspoiled coastlines. On the beach, bright lights and noise disturb and disorientate nesting turtles. Clearing forests has led to erosion, and the loss of mangroves has destroyed the breeding grounds of many reef fish.

Few hotels are connected to mains sewerage systems, and some pipe their sewage, untreated, into the sea, where their own guests are swimming. Many hotels also put large quantities of fertilizers and pesticides on to their lawns, gardens, and golf courses, but these, along with the many chemicals from cleaning and from swimming pools, can all leach into the sea.

Cruise ships have long been considered the scourge of the Caribbean, for the vast quantities of waste they produce, often poured or dumped in the ocean. When at anchor, their anchors and chains tear up great swathes of the sea floor.

But concerns among environmentalists and customers have begun to sink in, and some hotels and cruise ships are beginning to clean up their acts.

√ **Avoid polluters.** It can be difficult to ascertain for sure, but try to find out whether your hotel treats its sewage and wastewater properly. Extensive fancy gardens and golf courses right on the sea front are often accompanied by heavy use of fertilizers and pesticides, polluting the waters offshore.

√ **Support green hotels.** Many hotels choose to market themselves on the basis of their environmental credentials. Beware "greenwash", but be prepared to support those who are clearly investing in environmental matters.

√ **Ask questions.** Find out more about the environmental impacts of your hotel when you get there. If there are clear problems ask to see the manager, and demand action and a discount.

Northern Caribbean

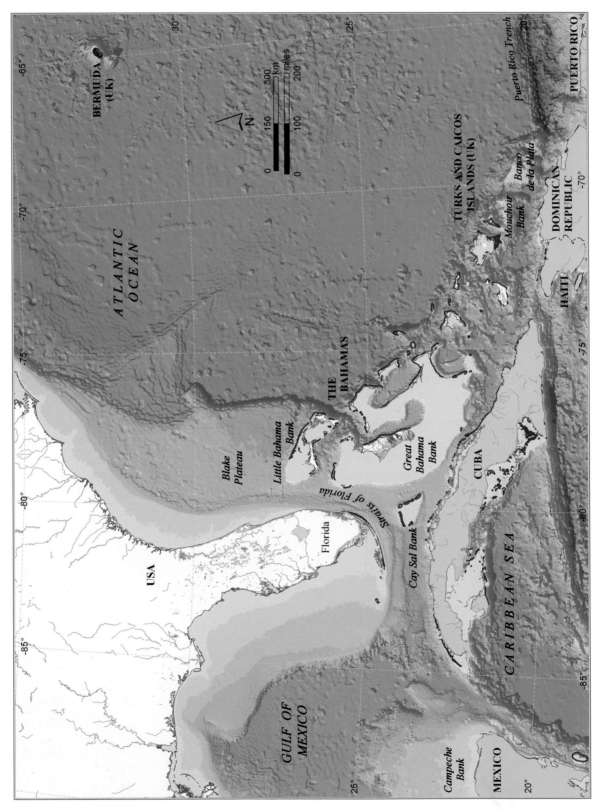

2.1 NORTHERN CARIBBEAN

Viewed from space on a cloudless day, the waters of the northern Caribbean are ablaze with a kaleidoscope of blues, splashed with dark greens and streaked with white. The wide shallow waters of the Bahamas and southern Florida are clearly visible in ultramarine, outlined by the cobalt blue of deep ocean. Coral reefs cross this region in a complex mesh, clinging to the outer edges of the shallow banks. Many are remote from human habitation, and to some degree still unexplored. Others, especially near the main population centers of southern Florida, Nassau, and Bermuda, are among the best known and most accessible coral reefs in the world.

Strictly speaking, these islands and reefs lie outside the Caribbean, making up the northern parts of what is called the Wider Caribbean. To the east, the tropical waters of the Antilles Current sweep past the outer edge of the Caribbean and the fringes of the Bahamas archipelago. To the west, warm water from the Gulf of Mexico spills through the Straits of Florida. These two great currents meet just to the north of the Bahamas and form the Gulf Stream, a great swathe of warm water which traverses the Atlantic Ocean and drives global weather patterns. They also carry warmth and life, in abundance, to the northernmost coral reefs in the world, at the tiny outpost of Bermuda.

These same currents are tracked and followed by one of the most magnificent beasts of the ocean, the humpback whale. Every winter these great whales leave the Arctic waters where they have been feeding and migrate south to mate and calve in the region of the Mouchoir and Silver Banks. They are regularly seen in parts of the Bahamas and the Turks and Caicos, and around Bermuda in April and May on their return journey.

The northern Caribbean includes vast areas of coral reefs, many of which are still in excellent health and offer fabulous opportunities to see reef life. While some have been damaged over recent decades, many appear to be recovering and efforts to protect them are increasing all the time. Visiting divers and snorkelers can expect to see the full range of Caribbean life in a tremendous diversity of locations. The Florida reefs provide something for everyone, but are perhaps typified by beautiful gardens of coral in calm waters.

Around the Bahamas and the Turks and Caicos, dramatic scenery is the norm with fantastic walls plunging into the deep ocean.

Bermuda must be the world's shipwreck capital, with some 400 wrecks spanning 500 years of history around this tiny island. Other reefs, far from land in the Gulf of Mexico, are difficult to reach but have fascinated scientists with their more unusual corals and other life forms. For those lucky enough to visit them, these are places of great beauty and remain home to some very large fish.

The Bahama Banks viewed from the window of the space shuttle. In the foreground are Crooked Island and Acklins Island with the Exuma Cays, the Tongue of the Ocean, Andros Island, and, only just visible, the Florida mainland. Cuba lies to the left of the picture.

Bermuda

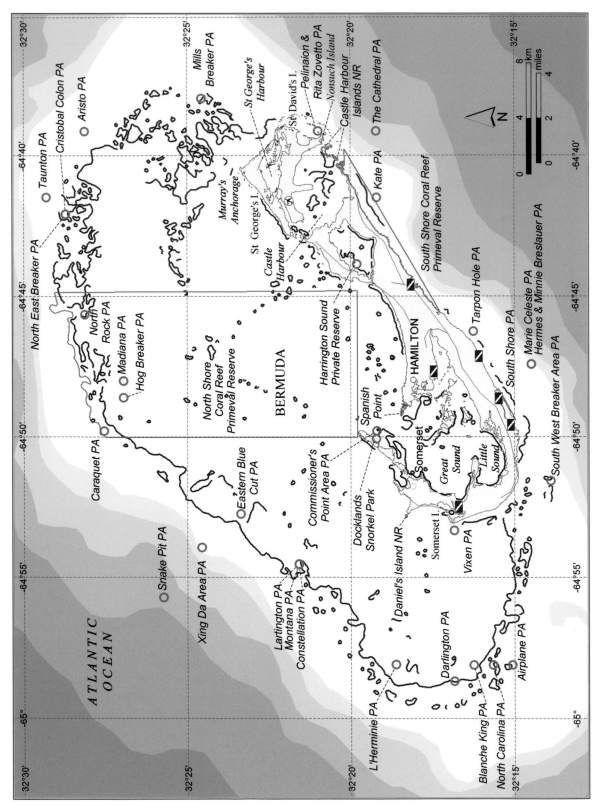

BERMUDA

Temperature	15-20°C (Feb-Mar), 25-30°C (Aug)
Rainfall	1 400 mm; no clear wet/dry season
Land area	39 km²
Sea area	450 000 km²
N° of islands	150
The reefs	370 km². Crystal clear waters and corals abound. There are fewer species here than elsewhere in the Caribbean because waters are cooler and the reefs are isolated, but these coral reefs are very healthy.
Tourism	278 000 visitors, plus a further 179 000 cruise ship passengers. Many come to snorkel and dive, but others are here largely for the beaches and golf.
Conservation	Bermuda has led the region in many ways, with turtle protection going back to 1620. Today there are several protected areas and fish stocks are benefiting.

Far out in the Atlantic Ocean lies a tiny patch of coral islands – little more than a speck on most maps. This is Bermuda, a British Overseas Territory. It lies at the same latitude as Savannah, Georgia, over 500 kilometers north of Florida's northernmost coral reefs – not a location you would expect to harbor corals, but even as you fly in to land it is plain that Bermuda is something special. Bright shimmering fields of turquoise and cobalt blue surround the islands, and the sea is alive with coral and tropical fish.

For those with an interest in the underwater world, two features make Bermuda a unique destination: crystal clear waters and a wealth of shipwrecks. On the outer reefs visibility reaches 80-90 meters. Under such conditions diving feels like flying, and with a backdrop of healthy reefs and countless fish this is an experience not to be missed. Although fish and coral diversity is lower here than in other parts of the Caribbean, the clear, clean waters around most of Bermuda provide the opportunity to experience a truly healthy and thriving marine environment.

These same reefs have presented a treacherous obstacle to shipping over the past 500 years. Some 400 wrecks are listed around the islands, representing the span of modern American history. Many are now little more than a pile of stones (once used as ship's ballast), or scattered cannons, but others offer spectacular

Black groupers are magnificent fish, quite commonly seen in Bermuda – the largest can be over a meter in length and may weigh up to 100 kilos.

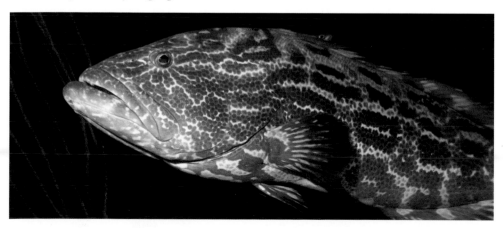

Tarpon are seen alone, or in small schools, over the reef or in mangrove areas. They are fairly inactive during the day, but hunt for other fish at night.

DIVING TO THE DEEPEST DEPTHS

The naturalist and explorer William Beebe chose Nonsuch Island in Bermuda as the first place in the world to begin to explore the deep ocean. His partner, Otis Barton, designed and funded the building of the Bathysphere, a tiny steel sphere, just large enough to hold two people. It had two windows through which they could observe their surroundings and was lowered on a steel cable, with a telephone connection to the surface and an electric light to enable them to see the world of the deep. Over a series of expeditions, the first in 1930, and with over 30 dives in all, they quickly overtook the greatest depth previously achieved by a human (160 meters) and eventually pushed the world record to over 900 meters (half a mile) in 1934.

Beebe's accounts of these descents give a tremendous sense of what this strange new world was like, including numerous luminescent creatures whose identification he could only guess at. "At 320 feet a lovely colony of siphonophores drifted past. At this level they appeared like spun glass... my next visitors were good-sized yellow-tails and two blue-banded jacks which examined me closely at 400 and 490 feet respectively. Here were so-called surface fish happy at 80 fathoms... At 800 feet we passed through a swarm of small beings, copepods, sagitta or arrow worms and every now and then a worm which was not a worm but a fish, one of the innumerable round-mouths or Cyclothones... At 2 000 feet I made careful count and found that there were never less than ten or more lights – pale yellow and pale bluish – in sight at any one time. Fifty feet below I saw another pyrotechnic network, this time, at a conservative estimate, covering an extent of two by three feet. I could trace mesh after mesh in the darkness, but could not even hazard a guess at the cause. It must be some invertebrate form of life, but so delicate and evanescent that its abyssal form is quite lost if ever we take it in our nets."

W. Beebe, *Half Mile Down*

structures among which the reef creatures jostle for space.

The islands themselves are today filled with an attractive array of low-rise buildings and verdant gardens. A few wild areas remain, including the Spittal Ponds Nature Reserve, which offers an important sanctuary for many wetland birds such as ducks, herons, and small waders. There are also a few stands of mangroves which, like the reefs, are the most northerly in the world.

Bermuda has a long history of environmental concern. In 1620 it enacted the first conservation law in the Americas, banning the killing of sea turtles. In more recent decades, concern for declining fish catches has led to much further legislation – the first protected areas were established in the 1960s and today there are a number of different reserves, including 29 dive sites which are completely protected from fishing. A result of these measures has been notable increases in fish populations.

History

Bermuda began life as a volcano in the mid-ocean some 110 million years ago. As volcanic activity ceased and the volcano itself began to sink back into the sea, corals and other creatures crowded its shores. They continued to grow, coral upon coral, and the base of the volcano is now smothered in limestone some 80 meters thick. Unseen by people, life evolved and flourished on this tiny coral outcrop.

The first human arrival was the Spaniard Juan de Bermúdez in 1503, and the islands of the "Bermudas" were marked on a map of 1511. One hundred years later, during a storm, the British ship, the *Sea Venture*, was badly damaged and separated from her fleet. Gradually taking on water she made her way to Bermuda, otherwise known as the Island of Devils, where she ran aground. The entire crew made it to safety and found a wonderful island with sufficient wildlife to feed them and sufficient wood to build two new ships. Nine months later most of the crew sailed on to America, but two of them remained to become Bermuda's first voluntary inhabitants. This story provided the inspiration for Shakespeare's *The Tempest*, which he completed in 1612. More settlers came, and over the next hundred years the island's economy grew. Alongside tobacco and a small whaling industry, piracy and wrecking were also the norm until Britain established a formal administration towards the end of the 17th century.

Bermuda has had colorful links with the USA. During the War of Independence a number of Bermudans stole gunpowder and gave it to General Washington in return for provisions. Later, America attacked Bermuda, only to be counter-attacked by an arriving British force. During the US civil war, Bermudan ships made a considerable profit running guns into the blockaded southern ports in return for cotton for the mills in Britain. The final turn of trade came during the US prohibition, when large quantities of alcohol were smuggled into the USA. Even Bermuda's role as a tourist destination for Americans began not for the white sand beaches but for the rum and whiskey.

TOUR OF THE REEFS

At almost 1 000 kilometers north of the tropics, it seems surprising that there can be so many tropical fish and corals in Bermuda. The reason for this – as well as for Bermuda's mild climate – is the Gulf Stream, which sweeps up from Florida and the Bahamas, bathing Bermuda in warm water and carrying life forms with it. Of course warm is a relative term; winter water temperatures can drop as low as 16°C, so good wet suits are needed.

Marine life begins right next to the shore, and many of the smaller beaches offer good snorkeling, particularly around the rocky headlands. There is even an underwater snorkeling trail, the Docklands Snorkel Park, which includes cannon from the 16th to the 18th centuries. In between these and other artefacts there is also a great range of reef life, with parrotfish, wrasses, tiny gobies, and blennies, and the occasional pair of butterflyfish. Across the broad shallow lagoon to the north of the islands numerous patch reefs offer pretty sights for visitors. Yellow pencil corals and green cactus corals are common, and the ivory bush coral, quite rare elsewhere in the Caribbean, is also widespread. Smaller fish abound.

The lagoon is edged by a shallow rim, 3-10 meters deep, beyond which a wide terrace harbors abundant corals. Breakers, or boilers, are widespread on the outer edge of the reef rim on all sides of the island. These rocks come close to the surface and are pounded by the waves during rough weather, creating spectacular plumes of spray. They are dislodged chunks of fossil reef, overgrown in parts by calcareous algae and tube-dwelling worms. On calm days, snorkeling or diving around the breakers offers spectacular scenery, with schools of larger fish including barracuda, as well as grunts, snappers, and hogfish closer to the bottom. Sheltered holes may be home to spiny and Spanish lobsters, and glassy sweepers often hover half hidden in the deeper recesses. North Rock, in the North Shore Coral Reef Preserve, is a popular site for snorkelers, while the nearby wreck of the *Cristóbal Colón*, a Spanish steamship which sank in 1936, has an impressive coverage of corals, including brain and star corals.

There are no elkhorn or staghorn corals anywhere in Bermuda, perhaps because the area is so isolated that the larvae have never reached it. The outer reefs are dominated by soft coral sea fans, sea plumes, and sea rods, while the most common stony corals are symmetrical and grooved brain corals, star corals, and mustard hill corals. Bluehead wrasse are common everywhere. Usually it is the smaller, bright yellow females that catch the eye, but the solitary blue-headed males are in many ways more beautiful. On all reefs these wrasses have favored spawning sites where the males take up residence for one to two hours every afternoon, and the females converge

The grooved brain coral with its distinctive valleys – which contain polyps – and deeper grooves between the ridges.

Silversides aggregate during the day around "boilers", shipwrecks, or other prominent outcrops. It is a remarkable sensation to swim through their midst.

to release their eggs. Wrasses live in a strict hierarchy and dominant males will chase off any intruders. However, if a site becomes vacant, it is quickly filled by one of the larger females. She immediately adopts the behavior of a male and within a few days will have undergone a complete sex change.

Larger fish, such as tarpon and black groupers, are common in many places, but one very special sight is the great aggregations of tiny silversides. The name silverside is used for several different types of small herring or anchovy species, though only experts can tell them apart underwater. They are nocturnal and during the day gather into large schools, often near caves, so densely packed they look like shimmering silver clouds. As you approach, the silversides, with an extraordinary singleness of movement, make a narrow pass or tunnel for you to swim through – you find yourself in a solid ball of fish, all just out of reach, but so densely packed you can see nothing else. These fish are popular food for many predators and it is common to see jacks or tarpon lurking alongside. There are few sights more dramatic than watching a hunt. The silversides move as one, in an instant making a tunnel as the predator races through, then massing together as a tight ball, or appearing to explode and swim in a thousand different directions. The purpose is clear: to confuse and disorientate the predator. But at dawn and dusk, when the light is fading and the silvery reflections on the fish are less dazzling, the predators, sometimes hunting as a team, make a catch. If you are there to see it you will never forget it.

Far off to the southeast of Bermuda (outside the area of the map) are two shallow seamounts, the Argus and Challenger Banks. Rising up from the deep ocean these undersea mountains come to within about 50 meters of the surface. This is certainly too deep for recreational diving, but in the clear waters of Bermuda there is still sufficient light for some corals to grow and there are reported to be true reef structures in places. A number of sport fishing boats go out to these banks but, like many other seamounts scattered through the world's oceans, they remain little known, perhaps teeming with life and with unique species, but too inaccessible for regular visits, even by scientists.

BALLOON-FREE NATION

Bermuda may be the only country in the world where releasing helium-filled balloons into the sky is an offense. Obscure as this may sound, there is a good reason for it. What goes up must come down and, in the case of a small island, any balloon released is highly likely to land in the ocean. Turtles often try to eat these balloons, which resemble tempting jellyfish. Every year hundreds, perhaps thousands, of turtles die from choking or starvation as a result of trying to eat garbage such as balloons or plastic bags floating in the sea. Banning the pointless and wasteful practice of setting balloons free is an inspired piece of legislation, putting Bermuda ahead of the game in trying to protect the natural environment.

UNITED STATES OF AMERICA

Temperature	Miami Beach: 17-23°C (Jan), 26-31°C (Aug)
Rainfall	1 128 mm (most of it in May-Oct); drier in Florida Keys
Land area	139 700 km^2
N° of islands	822 (Florida Keys)
The reefs	1 250 km^2. Tracing a long line from the mainland coast to the remote coral cays of the Dry Tortugas, the Florida Reef Tract is one of the largest and best known reefs in the Caribbean, with a great diversity of life. Other US reefs in the Gulf of Mexico are far from land. They are less diverse, but are generally in excellent condition, with very large fish and healthy corals.
Tourism	The number one industry. Two thirds of visitors to the Florida Keys take part in snorkeling, diving, or fishing. A large number of locals are also regular visitors to the reefs. There are more than 200 dive centers in southern Florida and the Florida Keys.
Conservation	Declines in corals and fish are being turned around in many places – fish stocks are recovering in a few important reserves, while there are also some significant areas of new coral growth.

Only a small area of the US coastline is warm enough to sustain coral reefs, and most of this is in the state of Florida. Florida stretches southwards in a great tongue of land from continental USA towards the Caribbean, and in the far south of this state lies one of the largest reef systems in the Caribbean, the Florida Reef Tract. This is also one of the best known and most visited coral reefs on Earth, offering unparalleled access to beautiful coral reefs for local inhabitants and millions of visitors. In contrast, there is a scattering of very poorly known coral reefs far offshore in the Gulf of Mexico, mostly in Florida, and two small but very special reefs off the coast of Texas.

Southern Florida is a land of contrasts. Physically it is flat – not a hill in sight – with shallow waters stretching to the south and west, edged by a beautiful chain of low islands, the Florida Keys. Most of the eastern shoreline and some of the Keys are now highly developed, with over 5 million residents in the zone stretching from Miami to Key West. In addition, tourists flock to the area for its climate, its amenities, and for diving, snorkeling, and fishing. On any day there may be between 500 000 and 800 000 visitors.

Tight against this development is the wilderness of the Florida Everglades – an enormous national park with a wealth of wildlife. Visitors should not miss any opportunity to experience the peace and silence of this land, and the thrill of seeing alligators and other creatures in their natural habitat.

The Florida Everglades is a vast, shallow, slow-flowing river, populated by alligators and a host of other creatures, and fringed by mangroves. The health of these ecosystems has a direct influence on the health of offshore coral reefs.

Lower Florida Keys

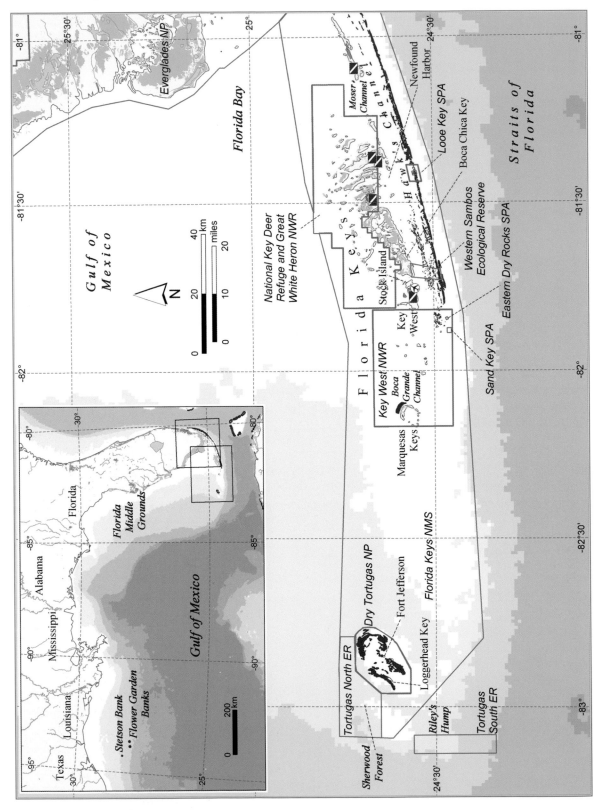

The shallow waters around the Everglades and the Florida Keys hold an abundance of ecosystems. Mangroves lie in wide forests on many coasts, an eerie sun-dappled world of water, mud, and trees, and a critical home for fish, crustaceans, and birds. Seagrasses carpet the sea floor over vast areas. These underwater meadows are populated by their own unique faunas, including the occasional grazing turtle, as well as the remarkable, enormous forms of the manatee (see box).

A brief history

The first peoples to arrive in south Florida can be traced back some 10 000 years. They were initially semi-nomadic, only later building permanent settlements. Little has come to light about their culture, but we know they traveled amongst the islands in dugout canoes and used the sea as a source of food. Remains of sharks, turtles, and conch, all caught and eaten by these people, have been found in ancient middens (waste dumps).

The Spaniard Juan Ponce de León found this land in 1513, but Europeans were slow to settle. By the late 18th century there were a few communities and missions and, as elsewhere, this led to the rapid demise of the native Americans, mostly through fighting and disease.

Florida was recognized as a state in 1845, and not long after this some of the first scientific work began in the area. Among the early pioneers was Louis Agassiz. His great study of 1849 showed, for the first time, the full extent of the Florida reefs, and proved beyond doubt that these terrible obstacles to shipping were in fact actively growing coral reefs. In 1880, Agassiz's son, Alexander, produced a map of the coral reefs and other habitats of the Dry Tortugas. This work was so detailed and accurate that modern scientists have used it as a base map to quantify changes in the extent of coral reefs and other organisms.

The 1880s saw the connection of Florida to the rest of the USA by rail, sparking a large influx of settlers and the beginning of tourism. The railway was then extended along the length of the Florida Keys in an environmentally damaging project between 1905 and 1912. A series of embankments and bridges was built between the islands, interrupting the water flow between the shallow Florida Bay and the nearby Atlantic Ocean, leading to great losses of seagrasses and fish stocks in the bay. A massive hurricane in 1935 destroyed the railway, but a road was built in its stead in the following years. The remote wilderness of the Florida Keys was open to development.

Florida's population grew rapidly throughout the 20th century, as much from immigration as from natural growth. Many older people retired to the warm climate, while others

FLORIDA'S MERMAIDS

West Indian manatees – possibly the origin of the mermaid legend – are extraordinary, captivating creatures. They are air-breathing mammals that have returned to the ocean. Looking like enormous seals, their forelegs have developed into short flippers and their hind legs have disappeared into a tail, but unlike seals they never leave the water. Although now rare, they are still to be found in a few places across the Caribbean.

Weighing up to 1 500 kilos and measuring up to 4 meters these gentle giants are entirely vegetarian, feeding on seagrass and algae. They are often found in small groups, but in the winter quite large aggregations gather in places where there is warmer water. They live up to 60 years. Mothers have a single calf which remains with them for up to two years, learning survival and travel skills.

Florida is a place of hope for the manatees. While in many parts of the Caribbean they have been decimated by hunting, here they are strictly protected. Some are still injured and killed when they are accidentally hit by boats, but today populations are increasing throughout the state. To the north there are a few places, such as Crystal Springs, where manatees have become major tourist attractions and where it is even possible to snorkel with them. Further south, travelers may still see manatees from kayak trips or boat tours amongst the mangroves and waterways of the Florida Keys, the bay, and the Everglades.

came to take advantage of the constantly growing tourism industry. In the latter half of the century the state became a focal point for Caribbean migrants and many regard Miami as the unofficial capital of the Caribbean.

Coral reef science

Many of the Caribbean's top coral reef scientists come to study at the universities and field stations of south Florida. Others come and never leave.

The first marine research station in the Americas was built on the remote Loggerhead Key in the Dry Tortugas in 1904 by the Carnegie Institution. This lab was operational

for 35 years, and provided a critical background to our knowledge of Caribbean coral reefs. Today, great centers for marine science include the Rosenstiel School of Marine and Atmospheric Sciences (RSMAS) in the University of Miami; the Tropical Marine Laboratory on Summerland Key in the Lower Keys; and the Florida Institute of Oceanography with its Keys Marine Laboratory on Long Key. Scientists have ongoing experiments all along the reefs and across the mangroves and seagrasses into Florida Bay.

One of the more intriguing field laboratories is the underwater Aquarius laboratory owned by the National Oceanic and Atmos-

South Florida mainland and Upper and Middle Keys

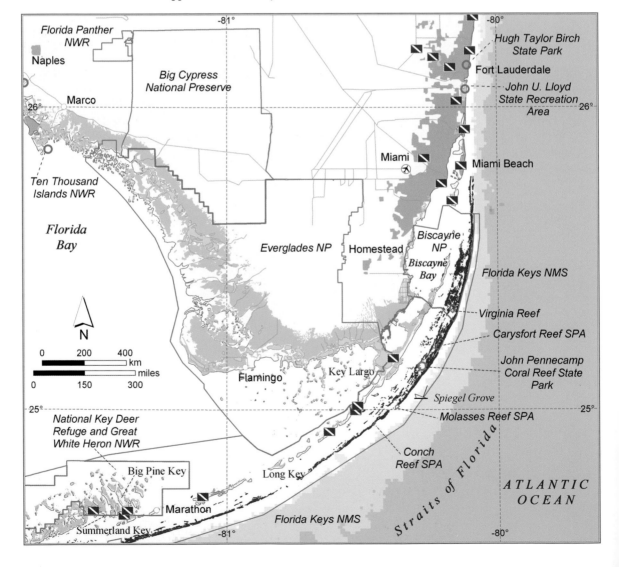

pheric Administration (NOAA) and run by the University of North Carolina. The lab is located almost 20 meters below the surface, but is equipped with computers, telephones, and access to the internet, as well as beds and cooking facilities. Scientists come and stay for about ten days – by living in the same pressure as the surrounding water there is no need for decompression and they can remain underwater making observations of the reef for hours on end. At the end of their stay, the aquanauts in the Aquarius are sealed in, and the pressure is gradually reduced over 17 hours so that they can decompress. In this way they can continue to work, eat, and sleep until it is safe for them to return to the surface.

Another laboratory has been installed by the Rosenstiel School on one of the cruise ships which regularly traverse the Caribbean. This fully automated laboratory passes oceanographic information back to the scientists in Miami from wherever the boat is located.

Elkhorn coral, once a major reef builder, is now rare, but still forms some beautiful gardens in Biscayne National Park.

Priceless wonders

South Florida is one of the best equipped areas in the world for enthusiasts to access coral reefs. Beginners can learn to snorkel and dive in safe calm waters; others can remain dry and see the reefs and other habitats from glass-bottomed boats. More experienced divers can dive deeper, visit wrecks, or take longer trips to the more remote reefs of the Dry Tortugas. Such interest is of immense value to the region's economy, and these are the most valuable coral reefs in the world. It has been estimated that they earn the region US$7.6 billion per year in reef-related tourism and fishing – that works out at about US$5 million for every square kilometer of coral reef every year.

Such interest has its costs, however. Tens of thousands of vessels traverse the waters and – while anchoring on coral is prohibited – many rip up the seagrass meadows with their anchors or the gouging of their propellers. Literally hundreds of vessels run aground every

THE STRANGE LIFE OF THE SEAHORSE

One Caribbean oddity sometimes seen in these waters is the seahorse. A keen eye and a degree of luck are needed to find them, but if you do, watch peacefully and they will not swim away. It is hard to believe that they are fish, but look closely and you will see a tiny transparent fin on their heads, just behind the gills, and another on the middle of their backs. They cannot swim fast, but rely on their near perfect camouflage to avoid being eaten as they wait patiently for their own food – small animals in the plankton – to swim past.

One of the many remarkable things about these strange creatures is that it is the males who give birth to young. When they mate, the female transfers her eggs to the male who then fertilizes them, and holds them in a pouch on his belly. Effectively, the male becomes pregnant and provides oxygen and some nutrients to the developing young. After a period of days or weeks the male goes into labor, a tough process which can last several hours.

Three species are to be found in Florida's waters, most typically in areas of seagrass and algae. One species, the dwarf seahorse, reaches a maximum length of only 2.5 centimeters, while the other two may reach 10-15 centimeters.

Longsnout seahorse.

The orange icing sponge lives in a tight relationship with some star corals. Growing together, each influences the growth and shape of the other.

the catches are reported to be greatest. Other large reserves include the popular Biscayne National Park, and two of the oldest marine protected areas in the region: the John Pennekamp Coral Reef State Park and the Dry Tortugas National Park.

TOUR OF THE REEFS
The Florida Keys

Florida's barrier reef is a vast structure that runs for 260 kilometers almost continuously from Biscayne Bay to Key West. In cross section it follows a common basic pattern along most of its length. In the shallow waters close to shore there are wide areas of seagrass, sand, and rock, but also quite a number of pretty patch reefs growing up from the bottom in about 5 meters of water. Further out the reef forms longer ridges parallel to the shore, with back reefs leading up to shallow reef flats and then falling away on the outer slopes where corals are most abundant. The outermost of these reefs are typically 5-10 kilometers from the shore and here the corals have built large spur and groove structures: high ridges of coral interspersed with deep channels. Deeper down, the intermediate and deep areas are often smooth or gently undulating, with extensive fields of coral.

The clarity of the water varies considerably from day to day and from reef to reef. As a general rule the inshore waters are murkier than those further from the shore, while those around the large islands are also a little clearer. Water running out from Florida Bay is the major cause of these differences and so the rise and fall of the tides also have a part to play, with waters usually at their clearest on rising tides.

Biscayne National Park

Just south of Miami, Biscayne Bay is a wide, mangrove-fringed bay. The shallow waters are dominated by seagrasses, with the reef marking the outer edge. Being so close to Miami, the park is very popular.

The reefs come up to within a meter of the surface in places and, with numerous shipwrecks, offer opportunities for snorkeling as well as diving. The reefs are not known for large fish,

year, each causing another small patch of damage to the ecosystem. Overfishing has greatly reduced fish stocks in many areas, while efforts to control fishing have often been met with fierce resistance. Harder when it comes to pinpointing blame, but equally damaging, have been the problems of pollution and the release of sediments which have taken place over many decades. Since the mid-1990s, much of the coral has died from disease and bleaching.

Thankfully, things are now beginning to change. The Florida Keys National Marine Sanctuary was established in 1990, and is one of the largest areas of protected reefs in the Wider Caribbean. Within its boundaries there are 24 fully protected zones (sanctuary preservation areas and ecological reserves) where there is strictly no fishing. Within just a few years of setting up these reserves, the numbers of fish have dramatically increased and the conflicts between fishers and divers have diminished. These zones have become the most popular dive sites, while fishers enthusiastically cast their lines just outside the reserve areas where

but are an excellent place for many colorful smaller species, including parrotfish, angelfish, squirrelfish, and damselfish. Stripy sergeant majors are abundant in the shallowest water, with blue and brown chromis a little deeper.

Elkhorn coral was once widespread here, and tended to be the dominant coral on the tops of the spur formations. Most has now gone, but in a few places, such as Virginia Reef to the south of the park, there are still wide patches of this strange, ragged coral. They give a hint of their former glory, offering a haunting and mysterious landscape where fish and lobster take shelter.

Upper Keys

Over 50 kilometers in length, Key Largo is the longest and most accessible of the Florida Keys, with the largest number of diving and fishing operations.

Carysfort Reef, to the north, is marked by a tall steel lighthouse that was mounted on the reef in 1852 (not before many vessels had foundered here, including HMS *Carysford* in 1770). Elkhorn and staghorn corals are coming back in a few places on this reef, while deeper down there are star corals and brain corals, with some complex scenery. Sometimes the larger parrotfish are to be found here, including the spectacular midnight parrotfish – a rich navy blue, with a scattering of paler, almost iridescent patches. Turtles – which it is actually an offense to harass in Florida's waters – are often seen in the keys, including loggerhead, green, and hawksbill. They are naturally inquisitive and a patient slow approach is often rewarded by some very close encounters.

Quite apart from the many ships that have foundered here over the centuries, a number have been deliberately sunk. At over 150 meters (510 feet), the *Spiegel Grove* is the largest of these. She was a military vessel built as a marine dock with a vast hold containing landing craft. This hold could be part-flooded to float the landing craft which could then motor out with hundreds of troops for beach assaults. In May 2002 she was sunk as an attraction for divers, and sits under 40 meters of water, rising up to within about 15 meters of the surface. Sponges and soft corals are already abundant on her sides, and a few large predators are to be found

lurking in her shadows. Over time, she will become more and more heavily encrusted with marine life.

Molasses Reef is another popular site. Here there are some centuries-old star corals reaching several meters in diameter. Many are twisted and undercut rather like ancient trees, and on some you will see the bright orange icing sponge. Quite a few sponges are capable of killing corals,

Ivory corals are typically found in quite small colonies on Caribbean reefs, usually below 25 meters. In the deep waters of the Oculina Banks, however, these same corals have built up high ridges, with individual corals reaching 1.5 meters across and almost 100 years old.

CORALS IN THE DEEP, DARK COLD

In 1975 scientists discovered some enormous banks of coral in deep (80-100 meters), relatively cool, waters near Cape Canaveral. Only one coral, the ivory tree coral, was found here and, although the same species can be seen in much shallower water across the Caribbean, at these depths this coral was a different color (white, rather than yellow-brown or pink-tinged) and lacked the algae normally found in its tissues. All around these banks were vast numbers of fish, and the site is a major spawning ground for the large gag grouper (popular with recreational fishers).

Similar large coral communities have been found even in very cold waters right across the Atlantic in recent years, but they are rare, and many have been damaged or destroyed by fishing boats dragging trawl nets across them. The Oculina Banks were given formal protection in 1985, but not before they too had been badly damaged by fishing nets. They will take many years to recover, but even what remains is beautiful and very important. It is also a reminder of how little we know about the underwater world, particularly when we look beyond the realm of scuba diving. How many more such places have been lost before we even knew of their existence?

but this species has developed a more compatible lifestyle. By covering the underside of the living coral they protect it from damage by organisms that might otherwise smother or drill through the coral. Molasses Reef is also famed for its fish – vast numbers swirl around the corals, including the ubiquitous grunts: French, tomtate, and bluestriped in particular. Bermuda chub are common and snapper are abundant in a few places.

Conch Reef has more dramatic scenery than most of Florida's reefs, with a short but steep wall, largely encrusted with sea fans and a great array of sponges including leathery barrel sponges and elephant ear sponges. Massive corals are abundant, including some spectacular pillar corals. Near this reef is an area closed to recreational divers where the submerged Aquarius research laboratory is located (see above).

Middle Keys

Well formed reefs with clear waters are not so numerous in the Middle Keys, but a few are renowned for their fish life. Southern stingrays are often found here resting on the bottom or

These fantastic structures are actually filters for sifting the water, and their owner is the Christmas tree worm. If disturbed they will retreat in an instant and cover the hole with a tight-fitting operculum.

occasionally feeding. From a distance you can watch these fascinating creatures without disturbing them. Divers may also occasionally be rewarded to see a yellow stingray. Although not as large or bulky as the southern stingray, these are beautiful creatures – dappled with a fine covering of yellow and black spots on a background of darker and lighter blotches. As with all stingrays, keep an eye out for the barbs on the tail.

The shallowest waters just above the reef are often covered in rubble, but behind these are fields of seagrass where keen-eyed divers may occasionally find a seahorse. There are often large numbers of queen conch here too. These giant snails with their magnificent winged shells are overfished almost everywhere in the Caribbean. Another giant, much more rarely observed, is the helmet shell. These hunt at night and, unlike their vegetarian conch cousins, they are predators, feeding on sea urchins which they are able to eat despite the spines.

Lower Keys

The famous Seven Mile Bridge takes traffic from the Middle Keys across the Moser Channel to the stretch of extensive islands known to many as the Lower Keys. In the east the National Key Deer Refuge is a wild area and the last remaining home of the key deer, a subspecies found nowhere else in the world. By 1957 fewer than 30 of these small, beautiful deer remained, but the population is now around 800.

A world away from these wild, mangrove-fringed islands is Key West, a vibrant urban center where tourism is the major industry and the holiday atmosphere has been perfected. Further west again, the Marquesas Keys are a scattering of ten uninhabited sandy and mangrove-covered islands.

Close to the main islands, the sanctuary preservation area of Newfound Harbor offers a great opportunity for snorkeling as well as diving. Being close to the coast, the water sometimes has a greenish tint, due to the murkier waters coming from Florida Bay, but there are spectacular numbers of fish. Thanks to strict protection from fishing it is also a good place to spot juvenile fish, many of which spend time here after migrating out from sheltered spots among the seagrass and mangroves.

Further offshore, many of the outer reefs have dramatic spur and groove formations on the outer reef slope, with high ridges dominated by stony corals such as star corals, starlet corals, and brain corals. Deeper down, these ridges become flatter and end in a relatively smooth slope, still covered in life. Looe Key is named after a British frigate, HMS *Looe*, which ran aground here in 1744, though only the experts can still point out the remnants of this boat. There are very large boulder corals here, and quantities of fish, including dense aggregations of snapper and grunt, regular visits by barracuda, and occasional schools of surgeonfish. Nassau grouper are often seen in the shadows.

A great strip of the reef running due south of Boca Chica Key is given extra protection in the Western Sambos Ecological Reserve. Since it was established, visitors have observed dramatic increases in the numbers of fish, spiny lobster, and conch. There are many spectacular large corals in the reserve area.

To the south of Key West and the Marquesas there are many opportunities for visitors to both snorkel and dive. Snorkeling is popular around Eastern Dry Rocks and Sand Key where fish are abundant, and will often come quite close. Be aware, however, that in many of the most popular snorkeling spots the corals are somewhat damaged. If you are a strong swimmer it is often worth swimming away from the boats and people to get a better impression of how the corals should look.

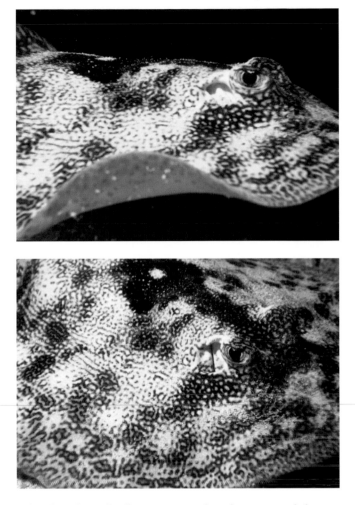

The relatively small yellow stingray, with its fantastic mottled coloration, is worth a closer look.

The Dry Tortugas

Far out to the west, the Dry Tortugas represent a marine wilderness quite unlike the Florida Keys. The strange-looking Fort Jefferson was built as a defensive structure on Garden Key during the US Civil War in 1846, although it never saw active service (except as a prison). Fort Jefferson and its surrounding waters were declared a national monument by President Roosevelt in 1935, making it one of the first marine protected areas in the Caribbean. Now called the Dry Tortugas National Park, this area is of critical importance for its wildlife as well as its cultural heritage. Over 300 species of fish have been seen in the park.

This is a place of very healthy coral reefs in crystal clear waters. Life abounds even on the outer walls of Fort Jefferson itself, making for some beautiful snorkeling. There are a few places where staghorn corals are now staging a comeback, but elkhorn is still rare. Further offshore there are a some large and impressive patch reefs. One such reef is Texas Rock, where there is a strange and spectacular seascape of dramatic pillars and mounds of corals, deep valleys, small caverns, and vertical walls. Giant corals, such as boulder brain corals, are plentiful. Large fish including groupers are common, and schools of horse-eye jacks are sometimes found here. Pelagic fish such as Spanish mackerel often swim in from the deep.

Recently, all fishing has been banned in the western half of the park, and commercial fishing is prohibited throughout the park. These measures were brought in because, although fish were still abundant in many areas, there were clear signs of decline. It had been shown, for example, that in 1930 the average size of a black grouper caught in the park was more than 10 kilos (22.5 lb), but by 2000 this had decreased to 4 kilos (9 lb). By halting the fishing, these species are being given a chance to recover and will begin to resupply other down-current areas of Florida's reefs to the east.

Beyond the park boundary, the Tortugas Ecological Reserve was declared in 2001. This reserve consists of two blocks that are strictly protected from all fishing and anchoring. Many of the reefs here are quite deep, but they include some very rich areas, such as Sherwood Forest, where corals are growing atop a fossil coral reef structure. Large cracks and hollows in the fossil reef reveal a deep sandy floor several meters below the reef surface.

The southern block of the ecological reserve, most of which is too deep for regular divers, is totally closed to visitors. At its northern end, Riley's Hump is critically important – thousands of groupers and snappers gather to spawn each year, and their eggs are swept away in the plankton and carried along the entire length of Florida's reefs.

Southeast Florida

On the mainland coast of southern Florida, northwards from Miami, small reefs and coral patches continue as far north as Vero Beach. Although not continuous, these reefs follow a clear pattern, with a shallow inner line, called First Reef, rising to within 3-4 meters of the sea surface. Second Reef forms a ridge in 6-8 meters of water, while the deepest Third Reef lies in 15-21 meters of water. In addition to these reefs, a large number of ships have been sunk along this coastline to act as artificial reefs for both fishers and divers.

First Reef is largely dead, a fossil reef left behind from a previous era, but there are some small corals, including the lesser starlet coral, and small mats of zoanthids. By contrast, Second Reef is covered in sea fans and other soft corals. In places there are also hard corals – star corals and the knobby brain coral. Large tracts of staghorn coral have recently been found in places on Second Reef and these appear to be growing vigorously, a real sign of hope for these

The barracuda can be an awesome companion to divers and snorkelers. Despite their appearance and inquisitive behavior, they are never a threat.

Soft corals tend to dominate the reef scenery of Florida, and in many places schoolmasters and grunts (here with a porkfish) drift in large numbers waiting for nightfall, when they move across the reef to feed.

now rare corals. Many of the same corals are found on Third Reef, but here there is an even greater diversity and the underwater topography is more dramatic.

The fact that reefs and fish are still found in these waters is a testament to their resilience. Onshore, the coastline is highly developed; pollution is a problem and even treated sewage threatens corals by adding too many nutrients to the waters. More threatening still is the widespread practice of "beach nourishment", which involves dredging sand from offshore areas and "feeding" it on to the beaches. Some reef areas have been severely damaged or killed by the shifting sand. Fishing is largely un-controlled along this coastline too, but, despite all of this, corals and smaller fish abound. Damselfish, angelfish, butterflyfish, and wrasse are dominant. In a few places, particularly amongst the stands of staghorn coral, there are schools of grunt and smaller snappers. The Third Reef lies close to the deep Atlantic waters, and is often swept by currents. Barracuda are widespread and amberjacks occasionally swim past. In a few places nurse, reef, blacktip, and even bull sharks may be seen.

Quite a number of dive operators take visit-ors out to these reefs, and it is worth choosing an operator with an interest in conservation – refraining from anchoring on the reefs and dis-couraging fishing or lobster collection.

The Florida Middle Grounds

Far away from the busy world of south Florida, and far from land, two long banks known as the Florida Middle Grounds have recently been explored in the Gulf of Mexico. The waters in this area are relatively cool and some tropical species are unable to survive, but others thrive far from any sources of pollution and away from the sometimes relentless pressure of fishing. The reefs rise up from the sea bed at 30 meters to about 21 meters below the surface, and are carpeted with sea fans and other soft corals, finger corals, elliptical star corals, and branching fire corals. Fish are numerous, including some large species such as coiba, amberjack, and big groupers, amongst them jewfish and gag.

Between these reefs and the Dry Tortugas a number of other submerged reefs have recently been described by scientists – and there may be more still awaiting discovery. Meanwhile, recrea-tional fishers, including divers with spearguns, are increasingly coming out to these places,

The Flower Garden Banks are one of the only coral reef areas with a permanent population of manta rays, whose strange horns (or cephalic lobes) help to direct plankton into their wide mouths.

Caribbean. The tops of the bank are bursting with a great abundance of hard corals, dominated by boulder star corals, with symmetrical brain corals, mustard hill corals, and great star corals. Many are eroded underneath, with the boulder-shaped corals often undercut to form mushroom-like structures. This is entirely natural, but is a potent reminder that corals, even massive stony boulders, are not permanent features of reefs. A host of creatures, including worms, mollusks, and sponges, as well as the crunching jaws of parrotfish, are quite capable of eroding, boring, and etching their way through even very large coral rocks.

The diversity of life on the banks is lower than on many more centrally located reefs in the Caribbean, probably due to their isolation. For example, there are no staghorn or elkhorn corals, sea fans, or whip corals. Small fish such as brown chromis and bluehead wrasse, as well as Spanish hogfish and reef butterflyfish, are common, but other groups, including grunts, are quite rare.

It is probably the large creatures for which the banks are most renowned. Some groupers use the banks as a spawning ground. A number of juvenile loggerhead turtles are resident here, feeding on a mixed diet of mollusks, crustaceans, and algae. Sharks, including hammerhead and silky, are abundant in the winter months.

Large rays are also common, and manta rays are present all year round. These ocean giants can reach nearly 7 meters across and weigh over 1 300 kilos. Their unmistakable characteristic is a pair of long flaps on either side of the mouth, usually coiled in a short spiral. These are used to direct water into their huge mouths, and the mantas then filter this water for their food, plankton. Mantas move with incredible grace through the water, often coming very close to divers.

The Flower Garden Banks were declared a national marine sanctuary in 1992, while a third bank, Stetson Bank, was added in 1996. The latter is not a true reef, although there are some corals, and it lies much closer to the mainland, where waters are more turbid and winter temperatures much cooler. Although anchoring and commercial fishing are now strictly prohibited on the banks, there is still some recreational fishing, and a small number of tourists also come out each year to dive.

finding an abundance of life long since gone from nearshore areas. The really big fish can be entirely removed from a reef in just a few dives and may take months or years to recover, so it seems likely that these reefs, still largely unstudied by the scientists, will be overfished within a few years.

The Flower Garden Banks

Almost 200 kilometers from land, and away from the other coral reefs in the Gulf of Mexico, two small seamounts rise to within 17 meters of the sea surface. Years ago fishermen noticed the bright colors of the sea floor in this area, and brought up corals and sponges on their lines and nets. It was they who gave the seamounts their name, the Flower Garden Banks. In this lonely spot conditions are just right for corals, with clear warm waters all year round, and plenty of sunlight filtering through on to a hard rocky floor.

Scientists have suggested that these may be the most pristine coral reefs in the entire

BAHAMAS

Temperature	Freeport: 17-23°C (Jan-Feb), 26-31°C (Jul); Mayaguana: 20-26°C (Jan-Mar), 25-31°C (Aug)
Rainfall	Freeport: 1 470 mm (most of it in May-Oct); Mayaguana: 850 mm (no clear wet season)
Land area	12 869 km^2
Sea area	652 000 km^2
N° of islands	1 891 (266 large)
The reefs	3 150 km^2. Mostly fringing and barrier reefs on the edge of the shallow banks, often linked to steep or vertical walls. Groupers and other large fish are still common. In a few places there are shallow coral gardens good for snorkeling.
Tourism	1.5 million visitors plus 2.5 million on cruise ships. There are 49 dive centers on 12 islands. The great majority of tourists go to Nassau and Freeport/Lucaya, so most other areas are uncrowded.
Conservation	Historically, many areas have been protected by their remote location. Today some species are declining, but there is a growing number of marine reserves where many species are abundant or increasing.

The Bahamas is still a relatively wild country. Scattered over a vast area of shallow sea it is made up of some 2 000 islands and islets. About 270 are large islands, but only 25 of these are inhabited.

The name Bahamas came from the early Spanish explorers who called it Baha Mar, meaning shallow sea. Around the islands lies a complex of shallow banks, with fringing mangrove forests giving way to sand, seagrass beds, and, of course, coral reefs. Sculpted by the tides and currents, these banks present a spectacular view from the air, with blues and turquoises extending for mile after mile. Very little of this country has been explored underwater.

Most of the Bahamas islands are found on two platforms, the Little Bahama Bank in the north and the Great Bahama Bank which makes up most of the center of the archipelago. Beyond these, the Cay Sal Bank lies off towards Cuba, while there are several other banks and islands to the southeast. These banks are built of limestone, some laid down by corals, but mostly formed from tiny granules called oolites that crystallized directly from the water. Over millions of years these deposits built up to incredible thickness – scientists have drilled through more than 5.4 kilometers of limestone.

The islands are flat or only gently hilly. During the last ice ages, the sea was much lower,

A yellowfin grouper haunts one of the deeper walls, with long rope sponges all around.

Northwestern Bahamas

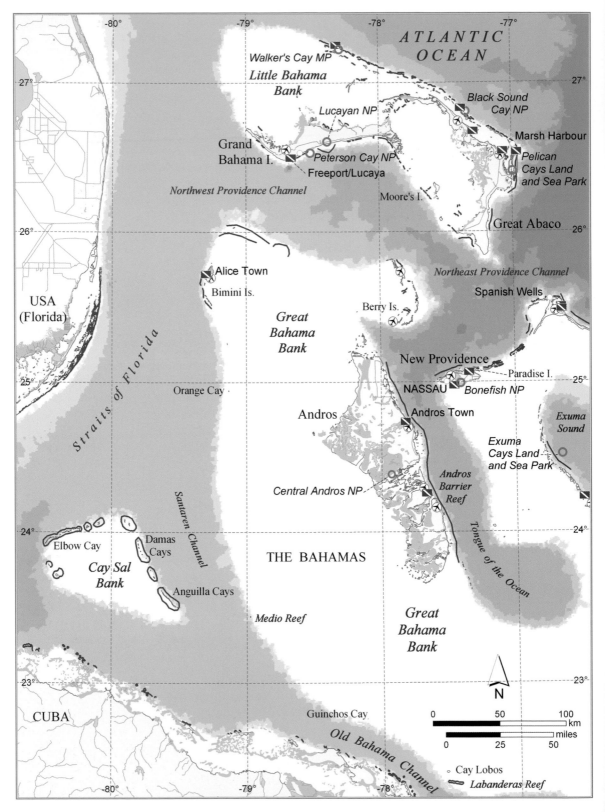

and the islands were vast, covering the entire area of the shallow banks. At this time the wind produced large sand dunes which later became rock when the sea eventually flooded back over the land. When the sea fell once more, the old sand dunes resulted in the more hilly parts of the islands we have today.

These same ice ages also shaped the islands in another way. All across the archipelago, and particularly around the island of Andros, there are spectacular "blue holes"– circular chasms of deep blue water cutting down through the shallow lagoons. They were formed when the banks were dry land, when underground streams dissolved away the limestone. Over time these streams carved out cave systems with occasional vast caverns beneath the surface. In places these underground caverns became just too big, and collapsed, leaving wide circular "sink holes". As the sea flooded the land again, at the end of the ice ages, these became the blue holes we see today. Expert divers have discovered complex underwater cave systems linked to the holes, but beware: diving in caves, especially in deep water, is extremely dangerous. Several divers have never returned from such explorations.

Coral reefs are mostly found on the very edges of the shallow banks, where the sea bottom plunges down dramatically into deep oceanic blue. Compared with many other parts of the Caribbean, these reefs are in good condition. Overfishing is not widespread, and visitors are entranced by the vast numbers of fish, including larger species. Dolphins are quite common among the islands and whales are also seen, especially during the migration of humpback whales between December and March.

Deep waters come in close to many of the reefs. One of the most impressive features of the central Bahamas seascape is the Tongue of the Ocean, a long sweep of oceanic water which wraps in around the north of New Providence and runs southwards along the eastern coast of Andros. Although quite narrow, the waters in this dead-end channel are over 2 000 meters deep. Many other channels cut between the banks, and their steep edges force powerful currents close to the coral reefs, bringing a good circulation of clean, clear water. In many of these places, huge schools of fish gather, and oceanic sharks and wahoo are regular visitors.

Seafaring peoples

Amerindian peoples inhabited the Bahamas for many centuries before the arrival of Europeans. The Arawak people, also referred to as Lucayans, arrived here around AD 600. As well as farming, they fished extensively, taking conch, fish, and even turtles and monk seals.

There were probably about 40 000 Arawak living in the islands when Columbus arrived on October 12, 1492, probably on the island of San Salvador. This was the first documented arrival of Europeans in the Americas, although it should be remembered that at first Columbus thought he was in Southeast Asia. In the following years many Arawak died from European diseases, or in resisting occupation. Thousands more were taken to work in the mines of Hispaniola, and within a generation they had virtually been wiped out.

Britain formally claimed the Bahamas in 1629, and in 1647 a group of British Puritans established a community in Cigatoo, which they renamed Eleuthera (after the Greek word for freedom). Other settlements followed, but the islands also became a base for pirates such as Sir Henry Morgan, Edward "Blackbeard" Teach, and Mary Read, who plundered the rich trade in vessels from the Americas to Europe.

Today, tourism is the most powerful sector of the islands' economy, although offshore banking makes an important contribution. Some 4 million tourists visit each year, mostly from the USA. This figure is over ten times the total Bahamian population. While some travel

LOST CITIES

For many years there have been rumors that Plato's lost city of Atlantis may lie in the Bahamas. Visitors to Bimini are still regularly taken to see the "Bimini Road", a strange formation of rectangular limestone rocks. Numerous other structures that look artificial have also been found in the shallow waters on the Great Bahama Bank. Most of this bank was dry land as recently as 8 000-10 000 years ago, and would have been quite habitable, but in reality we know that the native Arawaks did not arrive until much more recently.

Detailed surveys long ago proved that the Bimini Road, and other such structures, are nothing more than naturally formed beach rock which often cracks into rectangular blocks in this manner. Despite this, the extraordinary shapes are still being found, and mysterious legends and stories are constantly surfacing from these remote regions.

Southeastern Bahamas

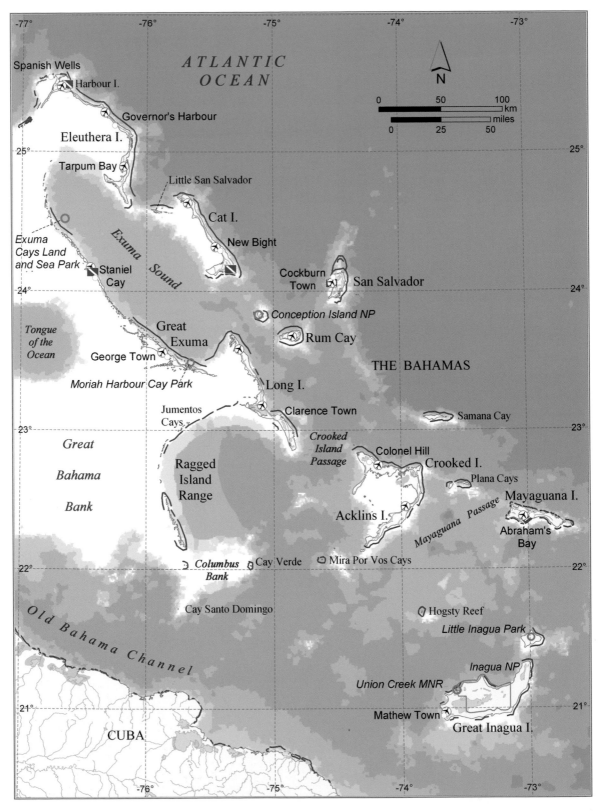

to the "out islands", the majority stay in the larger resort areas, and the country has two of the most highly developed tourist centers in the region – Nassau and Freeport/Lucaya – with hotels, casinos, and shopping malls.

While most tourists come for the climate and the beaches, many come to see the reefs, and even those who do not dive regularly often take an introductory dive or snorkel, or join a glass-bottomed boat trip. The Bahamas is also a top destination for game fishers, taking marlin in the deeper water and bonefish on the shallow banks. Quite a few operators encourage game fishers to release their catches, or at least those portions of the catch which are not going to be eaten, and where this is done game fishing can be quite compatible with conservation.

Protecting the oceans

There are now many marine parks and reserves across the Bahamas. In 2000 the government vowed to extend protection to 20 percent of the country's coastal waters. A first step in this direction took place in 2002 when a host of new marine parks was declared, including Central Andros, Little Inagua, Moriah Harbour Cay, and Walker's Cay. An inspiration for these efforts was the success of one of the oldest parks, the Exuma Cays Land and Sea Park. Continued protection of this site for many years has produced tangible economic benefits from tourism and improved fishing in nearby waters.

Fishing remains an important part of the economy, with large exports to the USA. The spiny lobster fishery is the fourth largest in the world, while the conch harvest is also important. In addition, the Bahamas has probably the largest Nassau grouper fishery in the world. Although this species is still widespread, numbers may be declining and these magnificent fish are still being taken at their annual spawning grounds, the worst possible place to fish if you want to be able to catch any next year.

Although turtles were once abundant throughout the islands, they are now rarely seen. Surprisingly then, local people are still allowed to catch turtles. There is a closed season and a minimum size at capture, but a significant harvest, amounting to several tons a year, is still taken, and turtle meat is still served in a few restaurants across the country. If you ever see

turtle on the menu do not order it, make a fuss, and walk out.

For a number of years the Tongue of the Ocean has been used by the US Navy to develop and test submarine technology and weapons out of sight of other navies. There have been some concerns about the impacts of these tests on marine life, and in 2000 the highly controversial Low Frequency Active Sonar was implicated in a mass stranding of whales and dolphins. This killed at least seven rare beaked whales, while none of the remaining beaked whale population has been seen since.

ISLAND TOUR
Little Bahama Bank

Large islands fringe much of the edge of the Little Bahama Bank and, behind these, the shallow waters are scattered with dozens of tiny cays and extensive areas of mangroves. These areas are important for many reef fish, such as grunts, which breed there or live as juveniles taking shelter amongst the mangrove roots and seagrass fronds. Numerous seabirds nest in the mangrove branches and feed in the shallows.

The Nassau grouper, now rare in many places, is still abundant across much of the Bahamas.

FORESTS OF THE SEA

Mangroves are found throughout the Caribbean. Often ignored, or regarded as smelly swamps, these are actually unique and beautiful places.

A visit to a mangrove forest must be planned properly. In some places it is possible to go by boat, or even rent a kayak, and there are also short boardwalks which enable visitors to walk in without getting wet or muddy. If none of these options is available then it is still possible to walk into a mangrove forest at low tide. Good shoes and insect repellent are recommended, and it is important to know when the tide will come back in!

As you walk between the mangroves and the sea you will see the extensive pencil-like roots of the black mangrove spiking straight up from the sand or mud, as well as the complex tangles of arching prop roots of the red mangrove. In some places it is quite easy to sink up to your knees in the fine mud. This mud is permanently waterlogged and so, because all plants need to get air to their roots, explains why mangroves have developed such strange forms. The pencil and prop roots all have tiny holes which act like snorkels to allow air in at low tide.

You are also likely to see the small fiddler crabs that make their homes here. The males have one very enlarged, bright orange claw which they use to signal both to other males (warning them to keep away) and to females (inviting them in).

One of the most exciting times to visit a mangrove forest is at high tide with a mask and snorkel, and the Bahamas is a great place to do this. You need to find relatively clear water, though it will never be as clear as on the reef. On the highest tides all of the roots, and even the lowest branches, become submerged, so snorkeling amongst them is like swimming in a jungle. Fish are abundant, with shimmering schools of juveniles and wary groups of snappers and grunts staring out from the protective shelter of the roots.

Mangroves are very important to marine life throughout the Caribbean. Many fish come in to feed, utilizing the shelter provided by the roots and foraging amongst the abundant leaf litter and debris. Others come here to breed. Many of the species seen on the coral reef began life in the mangrove forest, making it critically important that we protect mangroves alongside coral reefs and other shallow marine habitats.

Grand Bahama

The town and commercial center of Freeport/Lucaya on Grand Bahama is second only to Nassau in terms of both population and visitor numbers. Being only 90 kilometers from Florida it is a popular destination, with hotels, casinos, and a wealth of shopping, alongside beaches. Quite a few visitors venture over the reefs, including first-time snorkelers and divers. Despite the development onshore, there is plenty of good diving to be had in the ocean, even just a few hundred meters away.

To the south of the island the sea floor drops steadily down towards the deep Northwest Providence Channel. There are corals all along this slope, with coral heads in the shallower water, and then well formed spur and groove formations a little deeper. There is some good snorkeling to be had around the tiny Peterson Cay National Park and towards the east at the Lucayan National Park. The latter also has an abundance of caves both above and below water, including the most extensive underwater cave system in the country, with over 10 kilometers of mapped passages.

Swimming with dolphins is almost guaranteed on Dolphin Flats (just offshore from Freeport/Lucaya), where semi-captive dolphins swim freely out from their enclosure to play with divers. Although considered captive, these

dolphins are actually free to leave at any time. Elsewhere, in a couple of more remote sites on the bank, resident pods of dolphins have also been habituated to humans and regularly enchant visiting divers and snorkelers. Two species of dolphin are found in the Bahamas: the Atlantic spotted dolphin and the bottlenose. Though quite similar in shape and size, up close it is usually possible to see the extensive covering of spots on adult spotted dolphins. They appear to have an inexhaustible capacity for play and will regularly swim over to race with boats, apparently enjoying the speed and surfing on the bow wave. They regularly leap clear of the water, particularly when they are feeding (they typically catch fish and squid, but also take crustaceans).

A quite different activity, but one that is highly recommended for the more adventurous, is to take kayaks and paddle into the calm waters of the mangroves which lie behind the islands. In this serene "other world" there are numerous birds such as herons, noddies, and even pelicans, as well as many crustaceans and fish.

The Abacos

The eastern side of the Little Bahama Bank is edged by the long island of Great Abaco, and by a host of tiny cays which run parallel to this island and continue in a long chain to the north. This eastern edge of the bank drops rapidly and oceanic fish and sharks are often found close to shore, making it a popular spot for sport fishers as well as divers. The Pelican Cays Land and Sea Park off the east coast of Great Abaco includes small islands, mangroves, and coral reefs, with abundant fish life. It also includes extensive underwater caves. In the south of the island, the forests of the Abaco National Park are home to the rare Bahama parrot.

At the western end of the Abaco islands lies the newly declared Walker's Cay Marine Park in the waters surrounding the diminutive Walker's Cay. Long famed for its spectacular underwater scenery this place is also home to one of the more dramatic shark feeding spectacles. Regularly, groups of more than 100 sharks, including Caribbean reef sharks, but also lemon, blacktip, bull, and even small tiger sharks, are drawn to this site (see page 73).

Great Bahama Bank
Nassau and New Providence

Tourism has a long history in Nassau, the capital of the Bahamas, with the first Cunard steamship docking in 1860. Now millions arrive every year from the enormous cruise ships that pull in almost daily. Central Nassau is dominated by colorful colonial buildings, with numerous shops, largely catering for tourists. Nearby, the little island once known as Hog Island has been renamed Paradise Island. Connected by a bridge from Nassau, it has become a luxury hotel city.

Three large margate hide in the shadows of the "Vulcan Bomber" a wreck built specially for the James Bond movie, **Thunderball.**

Reefs are found along most of the northern shore, and along the small chain of islands which extends eastwards from here. Some of the best known reefs, however, have developed towards the southwest, and a number of dive operators run buses from Nassau to their bases on this coast. Here, from a depth of only about 10 meters, the sea floor falls away in a steep to nearly vertical wall that marks the edge of the Tongue of the Ocean. The top of this wall, festooned with coral, is constantly bathed in clear, clean water, and offers unique opportunities for "wall diving" in safe, shallow conditions. Above the wall are beautiful gardens of sea plumes and other corals.

Wrecks abound in these waters, and several have even featured in movies. Almost all have been deliberately sunk for the tourists, and the older wrecks are covered in life. The "Vulcan

Stunning scenes such as this, with sea rods, sea plumes, and fan corals, are common above the reef wall. Here a small group of female bluehead wrasse is seen in the foreground.

Bomber" (in fact just a tangle of scaffolding) from the James Bond movie, *Thunderball*, is now smothered in a fantastic profusion of soft corals, fan corals, sponges, tubeworms, and hydroids. Larger fish, snappers or groupers, often take up residence and can be seen in the darker recesses of many wrecks, peering out at passing divers.

The Bimini Islands

This string of tiny islands is the closest point to the USA, being only 79 kilometers from Miami, but between the two lies a deep channel, the Straits of Florida, which is constantly swept by the flow of the Gulf Stream. The bank edge here slopes down steadily and coral outcrops are widespread. This is one of the best places in the Bahamas to see large schools of snapper and grunts. Spotted eagle rays are often found here and nurse sharks are common, along with occasional oceanic visitors such as silky sharks and hammerheads. Game fishing, particularly for marlin, is also a major activity. The abundance of fish may be linked to the strong currents in the deeper water, but it is also a sign that overfishing is not a major problem.

Andros

Just 15 minutes' flight from New Providence, Andros is another world. It has the largest land area of the Bahamas, but is actually made up of three main islands and a host of smaller ones. Few people live here and large forests of Andros pine predominate. Vast areas are covered in mangrove swamps, criss-crossed with channels and occasionally pitted with blue holes.

Running along the entire length of the east coast, but often several kilometers offshore, is a bank barrier reef some 170 kilometers in length, one of the longest in the Caribbean. In general the coastline on this side of the island remains shallow, dominated by sand and seagrass, for some distance from the shore. Towards the edge of the shelf the water begins to fall away and corals become common. There are large areas of branching elkhorn and staghorn, with more extensive fields of corals, sponges, and some algae in deeper waters. In places these are highly sculpted into gardens, with pillars of coral and sponge-covered rock interspersed with sandy patches. Butterflyfish are regularly seen making their way through this seascape – usually in pairs as most butterflyfish mate for life. If a pair becomes separated as they move across the reef, one or both will swim upwards to get a better vantage point and so find one another again. At about 30 meters' depth the reef gives way to a vertical wall descending down into the Tongue of the Ocean.

The channels and mangrove forests of Andros present a critical habit for many species, including bonefish, making the island a top destination for fly fishing (see page 124).

Recognizing the importance of the mainly undamaged ecosystems of Andros, large areas of the island have been protected and there are efforts underway to expand the marine park system around the island.

To the north of Andros lies the small cluster of mostly uninhabited Berry Islands. There is a broad stretch of shallow sandbank to the east of these islands, dropping away to the inky depths of the Tongue of the Ocean, but to the south there are some spectacular coral gardens and steep walls, covered in life.

Eleuthera and Cat Islands

This whole region is lightly populated, and although levels of tourism are increasing, there is still not much pressure on the coral reefs. It is possible to find spectacular schools of snappers and grunts, and in places sharks, wahoo, and eagle rays are seen on almost every visit. A quick look at a map shows that there is a lot of reef here, most of it unexplored.

The longest and thinnest of the islands, Eleuthera is over 160 kilometers long, but only reaches 5 kilometers in width. Cat Island is similarly long and thin. Both face out on to the depths of the Atlantic and can bear the full brunt

SHARKS, SHARKS, SHARKS

Most experienced divers and snorkelers look forward to seeing sharks. They are beautiful creatures, powerful and sleek, their streamlined forms moving effortlessly through the water. Typically they inspire awe and respect, not fear. Sadly, though, seeing sharks is a rare event in most places, and even then they are wary, swimming past briefly before disappearing into the blue.

To be able to see sharks up close is rarer still. In a few places in the Bahamas, however, sharks have been regularly fed by divers, and have grown unafraid of humans. They have taken up residence (in New Providence, Andros, Walker's Cay, Long Island, and Grand Bahama) and can be seen even when there is no formal "feeding".

Some groups have expressed concerns about shark feeding. Feeding fish in general on coral reefs has been criticized because it alters their natural behavior, and may have wider ecological impacts. With sharks, however, the concerns are also for people. Feeding may reduce the natural wariness that most sharks have of humans, and also create an association between people and food. Sharks may therefore be more likely to attack, although to date there is no firm evidence of this.

In a counter-argument, sharks have been shown to be one of the biggest attractions for divers, with about 40 percent of divers placing sharks as a primary reason for choosing to visit the Bahamas. However, these visitors are unlikely to see sharks unless they go to a feeding site.

One study tried to calculate how much the sharks were worth to the economy of the Bahamas, looking at how much people were prepared to pay to see them, and estimating how many people would have visited a different country if it were not for the sharks. They concluded that each resident shark was worth US$15 000 per year – whereas they might fetch only US$15-70 when caught, killed, and sold for food.

Despite this value, the future for Bahamas' sharks is still not secure. Legal or illegal fishing could destroy their populations in a very short space of time. Dive operators and conservation organizations have been running concerted campaigns to demand more marine parks, not just to protect the coral reefs, but also the precious sharks.

It is important to remember that shark feeding is an unnatural experience. Divers form a large crowd sitting on the bottom and watch professionals draw in the sharks. It has more to do with Disneyland than everyday life on the reef. On the other hand, such visits present a unique opportunity to observe and admire these wonderful creatures.

Caribbean reef shark.

Older wrecks are easily recognized by the thick incrustations of life, like this dazzling azure vase sponge.

of Atlantic waves, giving rise to regular spur and groove formations. The channels are filled with sand and rubble that move back and forth with the surging waters. Rising above them are ridges of rock, often covered in branching and plate corals, and algae. Large fish such as horse-eye jacks and eagle rays are often found here, and it is a beautiful sight to look along a deep groove out towards the dark blue and see the sharp outlines of predatory fish hanging in the water. Sometimes the larger schools of jacks will come up and swirl around a diver in a tight circle. The reason for this behavior is still poorly understood, but one theory is that the fish are attracted by the divers' bubbles, which sparkle and vibrate in the water rather like small fish as they float up to the surface.

The steep wall of Exuma Sound comes close to the land at the southern edge of both Eleuthera and Cat Islands. This submerged cliff is smothered in soft corals, whip corals, sponges, and the dark bushy forms of black corals.

The Exumas

The western side of the Exuma Sound is lined by a string of tiny islets, the Exuma Cays. Most are uninhabited, although there is some development on the larger island of Great Exuma. Like the Tongue of the Ocean, Exuma Sound is a very deep channel cutting in to the Great Bahama Bank. Lining the edge of the sound are countless islands, all fringed by fabulous, largely pristine, and unexplored reefs.

The most extensive area of protected coral reefs in the Bahamas is here, the Exuma Cays Land and Sea Park, incorporating ten islands, plus many smaller islets and a large area of shallow bank and reefs. It was set up in 1958, and was declared a no-take zone in 1986. Policed by a warden and with help from visiting yachts and others, there has been no fishing in this area since that time. Today there are 31 times more conch in the park than outside it, and scientists have shown that conch, grouper, and lobsters are actually spreading out from the park to surrounding areas. Obviously this is good news for the fishermen nearby as well as for marine life.

Long Island, Jumentos Cays, and the Ragged Islands

The southernmost islands of the Great Bahama Bank are relatively little explored. Like the islands to the north, Long Island is quite exposed to the Atlantic swells. Dean's Blue Hole near Clarence Town is one of the largest blue holes in the country, and perhaps the deepest in the world. Visibility can be exceptional, but the waters descend to around 200 meters, so divers cannot even get close to the bottom.

The long, scattered chain of the Jumentos Cays and the Ragged Islands are more sheltered, and here once again there are miles and miles of coral gardens in the shallow waters, edged by coral heads dropping down in steep slopes towards a deeper wall.

The best known parts of the Bahamas, and particularly the Great Bahama Bank, lie close to the islands, but further afield there are wide areas still awaiting exploration. A number of dive operators have begun taking visitors to tiny islets or to coral reefs far from land, including Orange Cay and Medio Reef on the western edge of the Bahama Bank, and Labanderas Reef, Cay Lobos, and Guinchos Cay to the south – all reported to contain healthy reefs and abundant marine life.

The southeastern islands and banks

San Salvador was the most likely first landing place of Christopher Columbus. The island is something of a mountain top, standing out on its own in crystal clear Atlantic waters, with the sea floor dropping away in precipitous walls on all sides. Leading down to the wall are long stretches of spur and groove formations, creating deep canyons as they approach the reef edge. The walls themselves are full of complex caves and overhangs. Hammerhead sharks are sometimes seen in the deep water offshore. Working from submarines, scientists have discovered the deepest populations of reef corals anywhere in the Atlantic on these walls, with sheet corals at 119 meters and great star corals at 113 meters. This is a clear sign of water clarity around the island, enabling sufficient sunlight to reach into these depths.

Rum Cay, Conception Island, and Samana Cay lie on separate banks, also rising up from very deep waters, with dramatic walls all around. The uninhabited Conception Island is a national park, protecting large numbers of birds and the nesting grounds of turtles. Green and loggerhead turtles are regularly encountered underwater by divers. Samana Cay is rarely visited, but surrounded by coral reef.

Crooked Island and Acklins Island have dramatic wall scenery, coming close to the islands in the north. Out to the west are some tiny islets called Mira Por Vos (named by Columbus and meaning "look out"). They are surrounded by sand, studded with coral heads and small stretches of reef.

Mayaguana lies out to the east, while to the south is the atoll-like formation of Hogsty Reef. This nearly hidden reef has wrecked many boats over the centuries. It also lies on the route of migrating humpback whales which pass by in the winter months.

The southernmost islands of the Inaguas lie on a separate bank, close to the Turks and Caicos. Great Inagua is the third largest island of the Bahamas, with a small human population, but with the highest numbers of West Indian pink flamingos anywhere. Some 60 000 of these magnificent birds nest in the Inagua

French angelfish are often unafraid of divers. They largely feed on sponges, which most other fish find unpalatable.

National Park. Also within this park is the Union Creek Reserve, where there is ongoing research on sea turtles, notably the green turtle, which is still found here in some numbers. Tagged green turtles from Inagua have been found as far away as Venezuela and Panama. Although considerably smaller, Little Inagua is the largest uninhabited island of the Bahamas, perhaps in the Caribbean. The island and its surrounding waters were declared a reserve in 2002.

Cay Sal Bank

Far to the southwest, Cay Sal Bank is the most remote and little known of the Bahama Banks. The few scattered islands have no permanent inhabitants, but are important nesting grounds for the loggerhead turtle. The edges of the bank drop dramatically into deep water and there are extensive reefs, as well as a number of blue holes. Nurse sharks are commonly seen resting on the bottom of these, or under overhangs.

TURKS AND CAICOS

Temperature	Grand Turk: 21-26°C (Jan), 26-31°C (Aug-Sep)
Rainfall	600 mm; wetter Sep-Dec
Land area	491 km^2
Sea area	153 000 km^2
N° of islands	87 (22 large)
The reefs	730 km^2. Reefs are found on the bank edges, forming spectacular walls which come in close to the islands. There are some patch reefs in shallow water. Groupers and other large fish are still common.
Tourism	165 000 visitors. There are 17 dive centers on four islands. Most of the permanent population is on Providenciales, which is also the largest tourist destination.
Conservation	Tourism and population growth have come late to these islands and many of the marine communities are in exceptional shape, although rapid development and overfishing are just beginning to have an impact.

The Turks and Caicos Islands are a British Overseas Territory and still maintain strong links with the UK. They lie at the southeastern end of the Bahamas, and share common geological origins with the rest of the archipelago. Here again are small, low-lying islands on wide shallow banks which end in vertiginous drops into the surrounding ocean. There are extensive areas of mangrove wilderness, and vast areas of near pristine coral reefs, teeming with life. Many believe that the Turks and Caicos offer some of the best diving in the Caribbean.

There are two main island groups and this division continues underwater. The Caicos Islands are found along the northern edge of the large Caicos Bank while, to the east, the Turks Islands are scattered right across the smaller Turks Bank. Moving even further east there is another shallow bank, the Mouchoir Bank, which, although it has no islands, is also scattered with coral reefs.

The islands themselves have low or undulating rocky scenery, with large amounts of dry scrub and cactus in the Turks Islands, and low trees in the Caicos Islands. To the south of the main Caicos Islands there is a large area of intertidal habitat that includes mangrove forests and wide areas of mudflats with just a few saltmarsh plants. These swamps make a vast wetland wilderness of international importance, where rare birds such as the West Indian whistling duck and the reddish egret can be found.

Although no-one seems quite sure, the islands were probably named after the Turk's head cactus which is particularly common on the drier Turks Islands. The cactus gets it name from the young growths which are red and cylindrical, not unlike the fez hats once worn in Turkey. Caicos may simply be derived from the word *caya*, or perhaps *caya hico* which meant string of islands in the Lucayan Indian language.

Horse-eye jacks are often seen over the reefs. Powerful predators, they often hunt in packs, attacking schooling fish from several directions and trying to break them into smaller, more vulnerable groups.

Shipwrecks and settlements

It is still a matter of debate as to whether the first European on the islands was Christopher Columbus or Ponce de León, but, as in the Bahamas, European arrival led to the rapid demise of the local Arawak people.

The islands and shallow waters soon found themselves in the path of a busy trade route between the Americas and Europe, and of course presented a major hazard to shipping. Many wrecks have been found, including a remarkable one discovered in the mid-1970s about 30 kilometers south of Providenciales – thought for a time to be Columbus' ship, *La Niña*. This is probably not the case, but the Molasses Reef Shipwreck is the oldest known European wreck in the Americas. She probably sank some time between 1510 and 1520, and excavations have recovered an anchor, cannons, tools, surgical instruments, parts of the hull, and even parts of the rigging, which can be seen in the museum on Grand Turk.

Major European settlement from Bermuda began in the 17th century. Salt production became a major industry, while attempts at cotton and sisal plantations were quickly thwarted by the dry climate and poor soil. After a series of Spanish, French, and English rulers, efforts to formally link the islands to the rest of the Bahamas failed and they became a British Crown Colony in 1962.

Today, tourism is the number one industry, with visitors attracted by a quieter pace of life, stunning white sand beaches, clean water, and fantastic diving. From small beginnings this has been growing at 15 percent per year in recent times. Financial services are another important sector of the economy, while fishing remains a major employer. The Turks and Caicos export very large volumes of conch and lobster, although catches have declined a little in recent years. Measures are being taken to reduce overfishing through closing certain areas and by prohibiting all fishing during certain seasons.

The Turks and Caicos islanders are well aware that the environment is their most precious asset, and a great deal is being done to protect the coral reefs and other natural resources. A system of national parks and nature reserves has been set up covering a large area of the reefs and islands. The parks nearest the main islands are actively policed, and most dive sites have mooring buoys to combat damage from anchors. At the same time, however, there are moves for massive development off East and South Caicos, including building cruise ship ports. If these go ahead they will certainly change the atmosphere of a large and wild part of these lovely islands and will put considerable additional pressure on the natural resources.

Spiny lobsters abound in the Turks and Caicos, over the lagoons and on the reef walls.

From turquoise to deep blue

From the shore, the waters of the banks offer a dazzling backdrop of blues and turquoises in almost every direction. These shallow waters are dominated by vast areas of seagrass and algae, interspersed with patches of bare sand. Such areas are of vital importance for conch, and as nursery grounds for young lobster. Corals are largely found at the edges of the bank, where they have built spectacular reefs, like ramparts, on every side. These reefs slope downwards to depths of 10-30 meters, at which point they break sharply into a wall that drops away rapidly into very deep water. Spur and groove structures are common on the top of the walls, making for dramatic canyon scenery. The walls themselves are smothered in life, and large pelagic fish, including sharks and manta rays, are often seen in the blue waters beyond. Usually the waters around the

Turks and Caicos

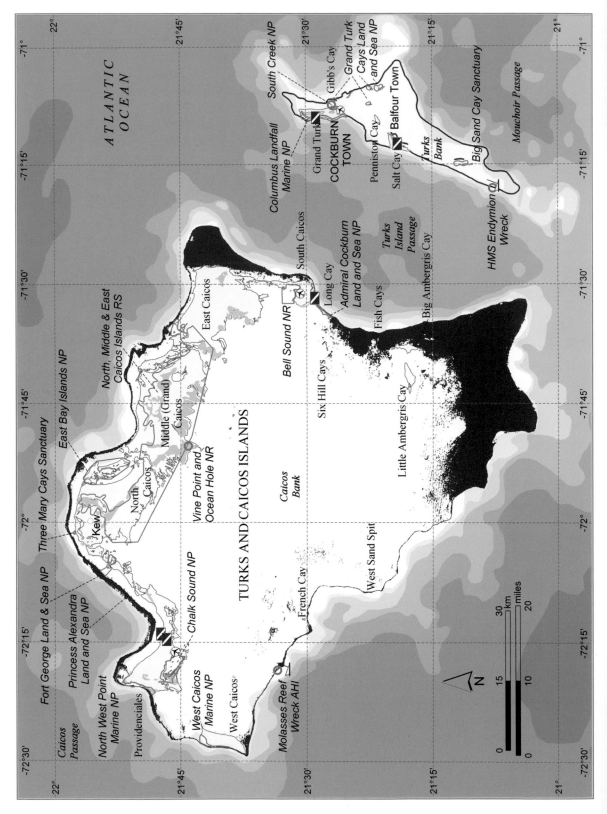

Turks and Caicos are crystal clear, but during a falling tide, particularly away from the islands, the water flowing off the banks can reduce the visibility, so it is a good idea to plan certain dives for a rising tide.

Although diving is a major attraction, there are also superb opportunities for snorkeling. Over areas of shallow coral gardens it is not unusual for snorkelers to get excellent views of loggerhead turtles, and even eagle rays and nurse sharks. Snorkeling amongst the mangroves also offers particular rewards, including young snappers and grunts sheltering amongst the roots, upside-down jellyfish, and even seahorses.

Between January and April, humpback whales can be seen as they migrate from the North Atlantic (see page 142). Many pass through the Turks Island Passage on their way to breeding grounds in the shallow banks nearby, including the Mouchoir Bank and the neighboring Silver Bank Sanctuary. In season, whale-watching trips are run from the Turks and Caicos, and this interest has also led to growing calls to protect the breeding whales and the Mouchoir Bank.

ISLAND TOUR
Providenciales and western Caicos

Providenciales, or Provo, is the largest population center of the islands and the main destination for visitors. A brilliant sandy beach runs along the waterfront, protected from the surf by a barrier reef to the north and providing sheltered waters for one of the island's most famous marine residents – Jo-Jo. In 1980 Jo-Jo, a wild bottlenose dolphin, left the company of other dolphins and began to seek out humans. In 1989 he was declared a national treasure and now has a full-time warden. Jo-Jo is not fed, but just seems to enjoy being with people, often approaching divers underwater. Two other dolphins are also sometimes seen with Jo-Jo and have been known to initiate games, picking up pieces of coral or sponge and passing them from one to another or to the watching divers.

Large stretches of this coast are protected, with the Fort George Land and Sea National Park to the east, the Princess Alexandra Land and Sea National Park along the northern shore, and the North West Point Marine National Park to the west. This protection has ensured that larger fish, including Nassau and tiger groupers, and even Caribbean reef sharks, are still widespread.

In general the reefs have the same basic structure all around the bank. The higher reaches of the wall are thickly encrusted with life and there are numerous black corals and wire corals. Wire corals are closely related to black, but never put out side branches and so grow longer rather than wider. The results are strange wire-like structures which often grow in loose spirals, pencil thin, but measuring up to 4 meters long. The wall is also a great place to see sponges: rope, giant barrel, elephant ear, azure vase, and even brown tube or moose antler sponges. There are many overhangs and small caves, as well as deep cuts and canyons leading down from the shallower water. Groupers and dog snappers can often be seen lurking in the shadows, with lobster in the smaller holes, and sometimes a resting nurse shark.

Just above the wall, in the brighter waters, soft corals are abundant and there are occasional high towers of pillar corals. Schools of grunt – typically French and bluestriped grunts, sometimes mixed with schoolmasters – spend their days resting tightly packed together amongst the corals, but spread out to hunt at night. More active during the day are the angelfish and butterflyfish, and the constantly busy bluehead

The diminutive slender filefish takes shelter amongst the fronds of a porous sea rod.

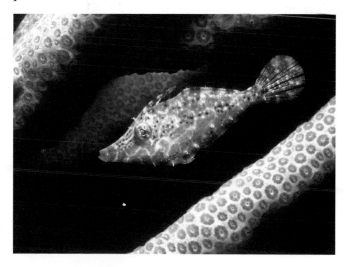

Wild dolphins, here a pair of spotted dolphins, are quite often seen underwater, and will approach divers and snorkelers in a few places in both the Bahamas and the Turks and Caicos.

wrasse and chromis. Loggerhead and hawksbill turtles are regular visitors to these coral gardens.

Towards the rocky shore of the uninhabited island of West Caicos there is some fabulous snorkeling and the water is always calm. Occasional visitors to these coral patches include schools of reef squid – usually just a few individuals, but sometimes there may be tens or even hundreds. Squid are highly intelligent and also very wary creatures, so the only way to get close to them is to keep very still, or to make only the gentlest movements – their own inquisitive nature can then sometimes draw them in. Like their relatives, the octopus, squid are masters of color change: at rest they may be a pale gray, perhaps mottled with white spots, but can change to dark brown in an instant. Even more impressive is when a school of these strange creatures changes color at exactly the same time. Such changes are a means of communication.

At night the reef seems quite different. Small creatures, including shrimps, bristle-worms, and basket stars, are regulars, but in many of the more remote locations night dives can be shared with horse-eye jacks and even reef sharks. Squid are often attracted to lights, allowing a very close approach, in contrast to their daytime behavior.

Further south the reef edge shows similar characteristics, although these areas can be subjected to much rougher water conditions on the surface. Around many of these reefs, away from the populated islands, Caribbean reef sharks are seen on almost every visit, with nurse sharks and eagle rays also making regular appearances.

Eastern Caicos

To the south of South Caicos lies the Admiral Cockburn Land and Sea National Park. The protection provided by this park has led to significant increases in both the size and numbers of fish on the reefs, and you are likely to see many groupers and snappers, while loggerhead and green turtles are abundant. South Caicos is home to the School for Field Studies, a training center for marine scientists. There are sometimes signs of ongoing experiments on the nearby reefs, such as settlement plates, which are simply tiles tied down to the reef and revisited over time to see how quickly corals and other creatures settle on them.

Eagle rays are numerous here. Like other rays, these magnificent creatures have powerful crushing jaws capable of breaking open even the thick shells of queen conch, one of their preferred foods. In fact this may explain why eagle rays are so often seen here. Part of the bank adjacent to South Caicos is a lobster and conch reserve and queen conch are very abundant. In the summer months, divers sometimes see schools of more than 100 of these rays swimming together, an unforgettable experience.

Turks Bank

Cockburn Town, on Grand Turk, is the capital of the islands, although in many ways it is a quieter place than Providenciales. It lies on the western edge of the Turks Bank, perched on the edge of an undersea wall that plunges down into the Turks Island Passage. The entire west coastline of the island has been declared a marine park – the Columbus Landfall Marine National Park – and conditions are typically calm as the prevailing winds are southeasterly.

The wall comes up to a shallow lip at about 10 meters' depth, and its dramatic scenery is enhanced by numerous canyons and undercuts, making an exciting playground for divers and a shelter for large fish such as grouper and tarpon. On the wall, deepwater sea fans form lace-like meshes more than a meter across. These lie like giant sieves across the flow of any current, and up close one can see the arrangement of polyps lined along the edges of the branches to maximize their ability to capture food from the plankton. Sea fans provide a popular resting place for the small, dappled brown form of the slender filefish. These fish are masters at camouflage, and readily change from pale gray to dark brown, adopting patterns of lines or blotches according to their background.

Rays are again common in this area. Eagle rays often swim past, while manta rays are regularly seen in the summer months. At one shallow site, just off Gibb's Cay to the southeast of Grand Turk, a number of southern stingrays have become accustomed to being fed by visiting tourists. In less than a meter of water these large rays will nudge and even clamber over snorkelers.

Gibb's Cay and a number of other northern cays make up the Grand Turk Cays Land and Sea National Park. Numerous seabirds, including spectacular colonies of sooty and noddy terns, are found around these islands. Ospreys are also seen out here. These beautiful fish-eating raptors can sometimes be observed flying over the shallow water and occasionally swooping down to capture a fish in their talons.

Further south still is a small settlement on the island of Salt Cay. To the northern edge of the island is a wide area of coral gardens offering fabulous opportunities for snorkeling amongst dense stands of elkhorn and pillar corals. Eagle rays, tarpon, and sometimes even hawksbill turtles are seen in this area. Porcupine fish may also be found sheltering in the darker recesses of the branching corals. Like the more common, and much smaller, balloonfish, porcupine fish are covered in short spines and can inflate themselves into large prickly balls if distressed or attacked.

Towards the very southern tip of the bank lie the remains of three wrecks dating back to the 18th and 19th centuries, notably the HMS *Endymion*. Although the ships are largely broken up, the bottom is littered with some 18 cannons and nine large anchors. It is fascinating to see how the reef has absorbed these wrecks, with many parts now only just discernible as the prolific growth of corals and sponges has smothered and changed the remains. Barracuda are a regular companion on these wrecks, watching their strange human visitors with interest.

Plumas are the commonest members of the porgy family in the Caribbean, often seen sifting through the sand for mollusks or other invertebrate food.

Mexico and Central America

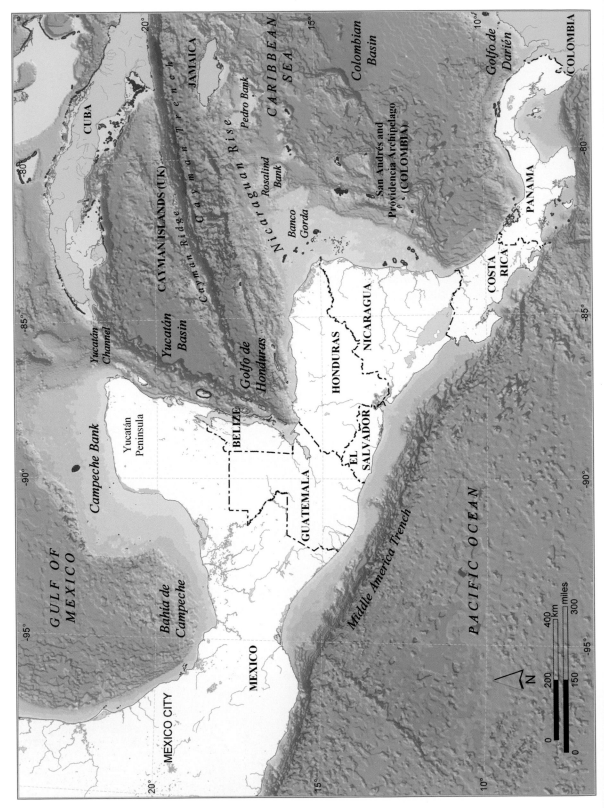

2.2 MEXICO AND CENTRAL AMERICA

The western boundary of the Wider Caribbean is marked by the long, snaking form of Central America, a land bridge between continents. Parts of this land are dominated by volcanoes, thrust upwards by powerful movements of the Pacific plate, but the Caribbean shores belie this drama. These are coastal lowlands, still thickly forested. Amerindian peoples – Maya, Miskito, Guamí, and Kuna – still live here, many fishing on the coral reefs as their ancestors have done for millennia. Rich coral reefs are widespread, though in places far offshore, away from the pernicious flows of muddy water from countless rivers. Coral atolls, barrier reefs, and island chains with thriving coral fringes, many still unsullied by human influence, are just now being explored by scientists and travelers. Closer in, dense mangrove forests and wide shallow shelves are home to crocodiles and manatees, while hordes of turtles come to feed and nest.

The reefs are home to some fascinating and unique wonders. On night dives, visitors to the reef may hear the extraordinary croaking calls of toadfish as they try to draw females to their nests. Satellite tracking has revealed a "local" population of whale sharks, the world's largest fish, and once a year these huge creatures gather off a remote reef in Belize to feast on the eggs of thousands of spawning snappers. Deep walls and spectacular scenery abound, and the diversity of corals is unrivaled. Recognizing the importance of these areas, wide swathes are being declared as national parks and nature reserves, and as tourism is bringing money and attention to these coasts so greater efforts are being made to secure the natural resources for future generations.

Only a scattering of corals are found in the Gulf of Mexico, most of these on the fringes of the Campeche Bank, but moving into the Caribbean Sea there is a transformation. From the Yucatán Peninsula to the Bay Islands of Honduras there is an almost uninterrupted sweep of fringing and barrier reefs, and offshore atolls. Scientists have begun to treat this as a single system, the Mesoamerican Reef. The connections between one reef and the next are considerable, especially thanks to the patterns of currents which swirl and eddy along this coast, and it is reasoned that the health of each reef will affect the health of the system. Great efforts are now underway to work together for its protection.

The Mosquito Coast, homeland of the Miskito people, runs from the eastern coasts of Honduras and down the wide coastal shelf of Nicaragua. Far offshore are uncounted numbers of coral reefs, still barely known to science, but of great importance to fishers. Moving across the borders from southern Costa Rica and into Panama healthy reefs are widespread in the area of Bocas del Toro. Then finally, in the far east of Panama, lies a truly unique place, the "land of the Kuna" or Kuna Yala – 365 islands and many more reefs in a long offshore chain. Here, perhaps more than anywhere else in the Caribbean, people are still an integral part of the coral reef ecosystem: a testament to the possibility of living in harmony with the world of the coral reef.

Guamí people live throughout the Archipiélago Bocas del Toro in Panama, close to reefs, mangroves, and the rainforest.

Western Mexico

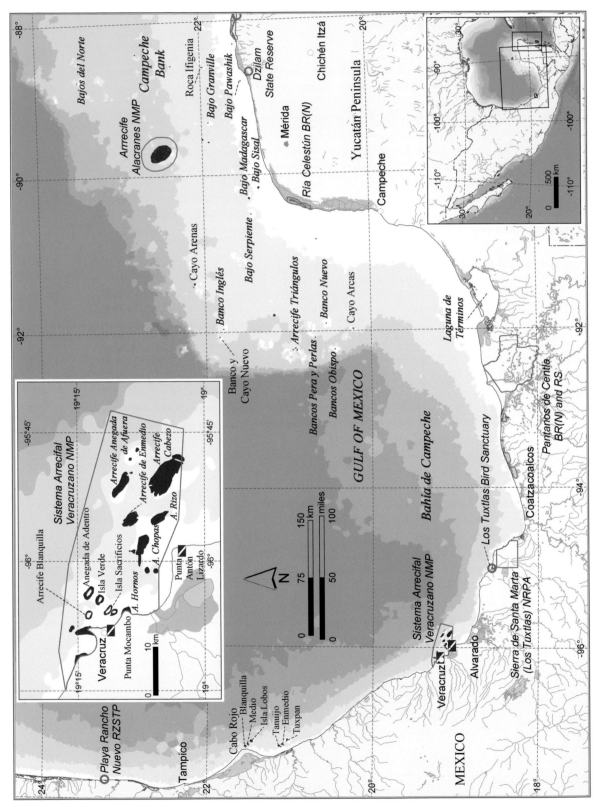

MEXICO

Temperature	Cancún: 19-27°C (Jan), 25-32°C (Jul-Aug)
Rainfall	1 580 mm; higher rainfall Jun-Oct
Land area	1 963 000 km²
Sea area	3 289 000 km²
The reefs	1 350 km². The eastern coastline of the Yucatán is packed with reefs, including a large coral atoll offshore; many are very healthy. Scattered patch reefs are found across the wide Campeche Bank and there are just a handful of reefs in the western Gulf of Mexico.
Tourism	2 419 000 international tourist arrivals, plus 1 595 000 cruise ship passengers. Vast numbers of these tourists go to the eastern Yucatán, many attracted by diving and snorkeling. There are 123 dive centers in eastern Yucatán, and six in the Gulf of Mexico.
Conservation	Tourism has transformed much of this coastline, but large areas have been declared protected. Differing levels of protection have been applied, but these sites will be increasingly important as tourism continues its impact on the rest of the coastline.

Mexico's lengthy Atlantic coastline runs in a great sweep along the western and southern margins of the Gulf of Mexico. It follows the Yucatán Peninsula then runs southwards again for a short stretch, facing east into the Caribbean Sea. Although this coastline lies within the realm of coral growth, the only extensive coral reefs are found in the east, across the Campeche Bank and into the Caribbean Sea.

The Yucatán Peninsula was the center of late Mayan civilization, from around AD 600. Moving north from their great cities in present-day Honduras and Nicaragua, these people developed new cities, such as Chichén Itzá, with vast complexes of stone buildings, sports arenas, and pyramid temples. They also used the coast, traveling, fishing, and even making pilgrimages by boat. Cozumel Island was an important religious center, with Mayan women regularly

The Mayan ruins at Tulúm are small, but among the most picturesque in Mexico. Here, as elsewhere, the Mayans lived in close association with the sea.

crossing over from the mainland by canoe to make offerings to the fertility goddess Ixchel. Remains of corals and shells have been found amongst burial offerings in the region.

CAVE DIVING

There are no rivers in the porous limestone of the eastern Yucatán, but scattered across the country are numerous sheer-sided holes dropping to neatly circular lakes. These holes, which were a critical water source for the Mayans, are known as cenotes (from the Mayan word for "well").

Geologically they were formed in the same way as the blue holes found in other parts of the Caribbean (see page 67), but have only become partly filled, and with freshwater rather than the sea. Today they are the center of some of the world's most dramatic cave diving. Ancient underwater caves lead off the cenotes, and in the 1980s expert cave divers began to explore these in earnest. They discovered a quite extraordinary world.

The water which fills the caves is of incredible clarity, and as the divers moved into these submerged catacombs they discovered mile after mile of vast, water-filled caverns and narrow tunnels, all fantastically decorated with stalactites and stalagmites.

Already more than 160 kilometers of passageways have been mapped. Connections have been made allowing divers to disappear in one cenote and re-emerge, miles away, in another. Some caves even connect to the sea, creating an odd world of shifting buoyancy and strange, oily mirages in the water where freshwater mixes with salt.

Cave divers must take considerable precautions – there is no "surface" to ascend to and, if they loose a light, become lost, or run out of air, they could quickly drown. With training, however, cave diving is a safe and indeed an exhilarating activity, and a few centers in the region now offer training, opening up an amazing world to experienced divers.

Enormous networks of caves are gradually being unraveled by expert cave divers beneath the limestone of the Yucatán. In places they open out into huge caverns, dripping with stalactites.

Although fighting with other Amerindian groups had reduced the size and strength of the Mayan cities by the time Spanish conquistadors arrived, they were still considerable. Cozumel, for example, had a population of 40 000. When Hernán Cortés arrived in Cozumel in 1519 he and his troops destroyed temples and religious artefacts, took slaves, and left an even more deadly legacy – European diseases, such as smallpox, which swept through the population. By 1600 there was no-one left on Cozumel.

Early European settlement was limited in this region, and to this day the Mayan people continue to dominate the population across wide areas. Over the last three decades, however, the land has been transformed. In the 1970s it was decided to promote the eastern coast as an international tourism destination, and now scores of cheap flights arrive daily from the USA.

Millions of visitors pour into the major tourist centers. Many are attracted by the climate and the beaches, but the underwater world has also been a major draw. Jacques Cousteau produced a documentary here in the 1960s, and further reports of deep walls, large fish, and healthy reefs have continued to enthrall and attract scuba divers ever since.

Scientific interest in Mexico's reefs began at around the same time as the start of the tourism boom, and there has been a burgeoning of research across the country. This has helped to plot changes and problems facing the reefs, and has also helped in the development of management measures to try to maintain these resources. The threats are very real – no hurricane could transform a coastline like the recent advance of mass tourism: with the changes have come pollution and disturbance to turtle nesting, as well as problems from the sheer weight of numbers of people wanting to get close to the coral reefs. To some degree these problems and threats have been recognized and there are ongoing efforts to tackle them.

A complex of protected areas now encompasses much of the coastline and coral reefs. Some of these are actively managed, and in Cozumel many of the dive shops collect a small fee from divers which is donated for the upkeep of the national park. In a landmark decision in 2001, the government overturned a permit to build a 1 400-room hotel in the Xcacel-Xcacelito

Marine Turtle Sanctuary. This was only after a sustained fight by environmental groups from Mexico and around the world, but perhaps represented a turning point. The protected areas have been strengthened, and environmental campaigners have been given hope that development will not always win out.

TOUR OF THE REEFS
Western Gulf of Mexico

A series of small reefs is found in the waters off Cabo Rojo, while a larger array is scattered in the waters off the major port of Veracruz, and off the small village of Antón Lizardo just to the south. These lie about 400 kilometers from their nearest neighboring reefs and lack the high levels of diversity of many Caribbean reefs, but are still beautiful places. Most are platform reefs, rising steeply up to the surface from 20-30 meters. Around Veracruz there are also a few, mostly degraded, fringing reefs. This city, and its associated heavy shipping activity, have created problems of pollution and overfishing, but the reefs become healthier, and the water clearer, further from shore.

The leeward (western) sides of the reefs have wide areas of tall sea rods and sea plumes, with large sponges and boulder star corals in the deeper water. The more wave-swept eastern margins have spur and groove formations, with quite a range of stony corals in the shallower water, notably finger corals, but also elkhorn, knobby brain, and fire corals. Star and brain corals become predominant on the deeper reef slopes. Sea urchins found here include the long-spined, slate pencil, and rock boring varieties. The last rest during the day in small holes which they have carved themselves with their teeth; they will return to the same hole each morning, defending it against any rivals.

Blue tang are abundant, bluestriped grunt rest out the day in dense schools, and angelfish – French, gray, and rock beauties – add a sense of majesty. Greater soapfish are also found, resting by day, lying partly on their sides (often mistakenly considered to be ill). These strange relatives of the groupers get their names from a soapy secretion on their skins which is a defense against predators – it is quite toxic and, on contact with human skin, forms a lather and brings out a rash.

Campeche Bank

The entire western and northern coastline of the Yucatán Peninsula is bounded by one of the broadest continental shelf areas in the Wider Caribbean. There are dozens of remote coral reefs, mostly on the outer edges of this shelf and up to 130 kilometers from shore. Growing up from depths of 40-60 meters, some are sub-

Greater soapfish are often seen reclining at strange angles on the reef slope during the day. At night they become animated predators, catching small fish, shrimps, and crabs.

merged banks, 10-15 meters below sea level, but most are shallow platforms, some marked with small coral cays. Some of the larger reefs, such as Arenas and Arcas, reach more than 3 kilometers in length, while Alacranes is 22 kilometers long, and has five small islands.

Scientists have drilled through these reefs and found that, geologically speaking, they are quite young. Indeed these may be the fastest growing coral reefs on record, with an upward growth of coral rock estimated at 12-14 meters per 1 000 years. The rich banks of staghorn coral which produced such growth are, at the present time, largely dead, but they may recover, and meanwhile there is an abundance of other corals.

Eastern Mexico

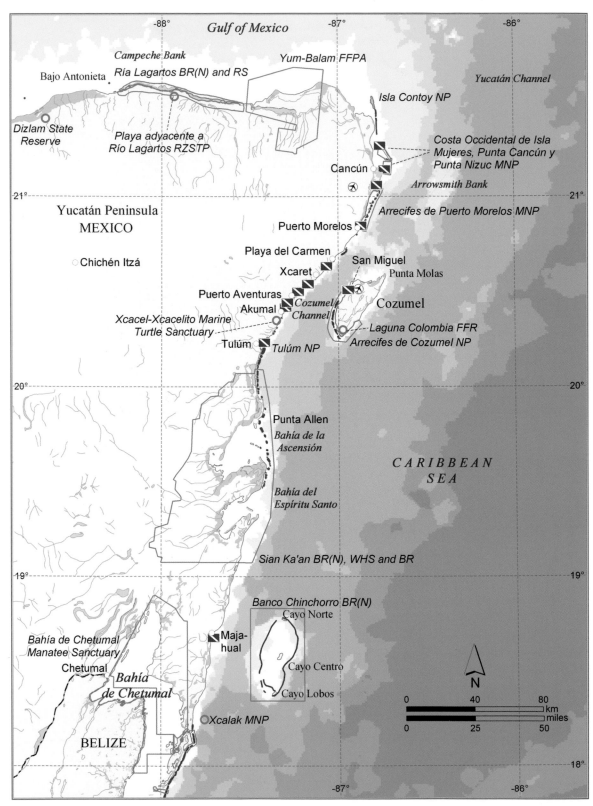

Most of the reefs are oval in shape, with a long, crescent-shaped, windward edge traced by breaking waves, and a more gradual slope to the southwest. The giant formation of Alacranes, which is beginning to attract small numbers of tourists, has a lagoon which reaches 20 meters deep, rather like a coral atoll. This shelters a complex network of inner reefs. The windward edge of the reefs drops quite rapidly.

In the wave-swept areas, colonial zoanthids are plentiful. These are close relatives of corals and sea anemones, with small, rubbery polyps with a wide, flat center ("oral disc"), fringed with a circle of tentacles. A little deeper some elkhorn corals still remain, along with extensive fire corals. Deeper still, growing on the high coral spurs, are wide areas dominated by lettuce coral and a variety of star corals. Large black and red groupers, for which there are important fisheries in these waters, may be seen at lower depths.

While most of the reefs are in good shape, there is a major oil pipeline and deepwater terminal near Cayo Arcas, with problems of oil and other pollution nearby. Overfishing is also encountered over the entire Campeche Bank.

Inshore, the Campeche coast of Yucatán has some important natural areas. The mangrove forests, especially around Laguna de Términos and Ría Celestún, are quite unlike the low mangroves more common on coral reef islands, with individual trees reaching heights of 20-30 meters. Turtles nest on many of the remote beaches.

Caribbean coast

The northeastern corner of the Yucatán Peninsula is a place of remarkable transition: as the coast swings abruptly around to the south we move from the Gulf of Mexico to the Caribbean Sea. The wide Campeche Bank gives way to deep waters coming close in to the continent, and coral reefs fringe almost the entire length of the coastline. The Yucatán Channel between Mexico and Cuba funnels the Caribbean Current into the Gulf of Mexico and there is a near continuous rush of water, the Yucatán Current, flowing northwards along this coast.

The coastal land, all within the state of Quintana Roo, is low-lying limestone, still dominated by forests and small-scale agriculture, but in a narrow fringe along the shore there has been an extraordinary explosion of tourism. Nowhere

is this more felt than in Cancún. In the 1970s this was an isolated sand bar lying offshore from a lonely, mangrove-fringed lagoon. Today there are high-rise hotels extending for more than 20 kilometers along a fine, white sand beach. Large shopping malls and international restaurants add to a rather bizarre other-worldliness, far, far away from the rest of Mexico.

Although this tourism boom has been more measured than elsewhere, even today the last pieces of real estate are being sold off, and will soon be filled with exclusive hotels, private condominiums, and dive centers. Thankfully, while this has been going on, the government has set about protecting other areas of coastline from the same development. There are some large national

Reefs lend flashes of color to the shallow waters south of Cancún.

parks and many smaller reserves where a flavor of the original and wild Yucatán can still be found, and where turtles can still haul themselves out of the water, unmolested, to lay their eggs.

Northern Caribbean coast

From Isla Contoy to Cancún the coastal shelf is still relatively wide, and there are sometimes upwellings of nutrient-rich waters from offshore. Here there are broad areas of sand and only scattered patch reefs, sometimes elongated into

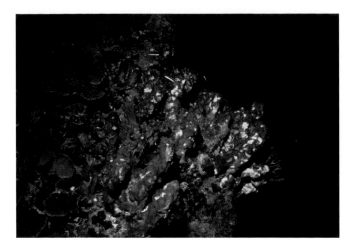

Sponges abound on Cozumel's famous southwestern reef walls.

ridges. Sea rods, common sea fans, and star corals are widespread, and elkhorn corals are making a recovery in some areas. It is the extraordinary abundance of fish, however, for which the reefs are renowned. The bright stripy forms of sergeant majors swirl around visitors, while dense packs of French grunt, cottonwick, and schoolmaster drift amongst the corals. Numerous

A southern stingray rests in a narrow sand channel on the lower reef slopes of Banco Chinchorro. These channels become part of the deep spur and groove formations in shallower water.

angelfish are found all along this coast, and it is a good place to see the rare blue angelfish – similar to the queen angel but not quite so bright, with less yellow on the tail, fins, and cheeks, and lacking the blue "crown" on the forehead.

Between the tourist centers of Cancún and Isla Mujeres lie some of the most heavily visited coral reefs in the world, where glass-bottomed boats and even a submarine join divers and snorkelers. There is inevitably some damage on the more popular reefs, caused by physical touching and breaking of corals. Further offshore, where there can be strong currents, the reefs are very healthy.

Unlike its neighbors, Isla Contoy is uninhabited. It is protected as an important seabird sanctuary, with brown pelicans, double-crested cormorants, and thousands of magnificent frigate birds. Green, hawksbill, and loggerhead turtles nest on its beaches.

Moving south, towards Puerto Morelos and beyond, the reefs are more continuous. Currents are a regular feature, particularly on the deeper reefs. These reach up to four knots in the center of the Yucatán Channel (double or treble most people's swimming speed). Large barrel sponges are common, and sea fans and sea whips reach up into the shifting waters to filter for food. Schools of blue tang rove over the reef and occasionally dive down to feast on algae, while porkfish add bright splashes of color to the scene. Over sandy areas peacock flounder are regularly observed, and may be seen changing color as they work to blend in with their surroundings. Smaller groupers are abundant everywhere, while larger yellowmouth and black grouper are found in deeper areas.

Some of the best formed reefs lie in the waters to the south of Puerto Aventuras, where there is a fore reef with spur and groove formations descending to 10 meters or so. Tall pillar corals rise up above the reefscape, and the ridged cactus coral is found in a few places.

Down amongst the corals are many of the small, cigar-shaped wrasses – the ubiquitous bluehead, but also yellowhead, clown wrasse, and slippery dick, with the occasional much larger, and shyer, puddingwife. One small fish, often overlooked, is the sharpnose puffer – a tiny but plump fish with a pale belly and bright bulging eyes – they get their names because, like balloonfish, they puff themselves up with water when threatened, to make themselves too large for predators to swallow. Loggerhead, green, and hawksbill turtles are regularly found in these waters (the name of the resort town of Akumal comes from the Mayan word meaning "place of the turtles"), but it takes careful observation to distinguish the different species.

Cozumel

Most visitors come to Cozumel for diving. On the leeward (western) side of the island the sea floor shelves gently down to about 20 meters, then plunges to 400 meters or so in the Cozumel Channel. In the south of the island, close to the coast, this inner shelf houses some especially pretty reefs, home to countless fish, but it is the shelf edge which has attracted most attention.

Off the southwest coast an ancient reef, no longer actively growing, marks the edge of the deeper water, and here the forces of erosion have sculpted an incredible seascape. Towering buttresses, carved through with caverns and long tunnels, rise out of bright sand arenas. Dizzying cliffs, split by deep fissures, fall away into deep ocean blue. Powerful currents sweep through this extraordinary world, carrying divers through astonishingly clear water.

There are few living corals here, but a great array of rope, tube, stovepipe, and barrel sponges deck the walls with red, yellow, brown, purple, and orange, with the occasional dazzling flash of azure vase sponges, while hanging vine algae add shaggy coats of green. These walls are busy with life. Arrow crabs and banded coral shrimps loiter in smaller crevices, and brittle stars smother some of the sponges. Bright purple and yellow fairy basslets swim over the walls, while the deep blue of the open water seems sometimes to boil with huge schools of creole wrasse and Bermuda chub. Large black groupers and jewfish also live here; stingrays rest on the sandy slopes, and sharks and turtles sometimes swim past.

Cozumel is regularly included in lists of the top dive sites of the world, but such an accolade comes at a price. These are among the most heavily dived places in the Caribbean. A rush hour of dive boats heads to and from the reefs morning and afternoon, and the more popular sites can be crowded with divers.

By contrast, the east coast of Cozumel is rarely visited. Waves crash on to this shore and the heavy swell, combined with currents, makes diving difficult. All along, however, are reef formations a little more typical of the region, with high spur and groove formations dropping down into deep water.

Southern Caribbean coast

Reefs continue across the mouth of the two great bays that form the heart of an extraordinary wetland wilderness, Sian Ka'an (Mayan for "origin of the sky"). This site is a mosaic of forests, grasslands, mangroves, sand dunes, and wide expanses of water. A vast range of birds is found – more than 340 species, including greater

THE SPLENDID TOADFISH

In the smaller recesses of Cozumel's reef can be found the splendid toadfish, a special fish which is unique to this area. With staring eyes, a broad mouth, and a straggly beard of fleshy barbels, this wonderful fish is marked with fine tan and gray lines across its head, blotched over the rest of its body, with smart yellow margins to its fins and tail.

But it is not just their appearance which makes members of the toadfish family remarkable. Male toadfish are strident vocalists. At night, sitting in "nests" or sheltered spots, they produce loud, deep croaking or belching noises, trying to entice visits from females. Divers can often locate them by following these sounds. If they manage to attract a mate, the same males will stoutly defend their clutch, and continue to protect the young toadfish after they hatch.

Unlike almost all other reef fish, young toadfish do not have a larval phase, and so they cannot be swept long distances to colonize new places. Over evolutionary time this means that many toadfish

have become isolated and have evolved into separate species. Several, like the splendid toadfish (pictured here), are restricted to single countries or islands – the whitelined and the whitespotted toadfish are found only in Belize; the cotuero toadfish only round the offshore islands of Venezuela. Cozumel's other resident toadfish, the large-eye toadfish, is more widespread, found as far south as Panama.

flamingos, jabiru storks, and roseate spoonbills. On land there are howler and spider monkeys, tapirs, jaguar, and deer, while in the waterways there are two species of crocodile, and manatees are still widespread.

Further south again the fringing reefs lie in the shadow of Banco Chinchorro; this area has

Groups of up to 20 spotted goatfish are often seen moving over the reef and using the strange "barbels" under their chins to sift through the sand in search of food. The fish on the left is a male yellowhead wrasse.

been spared hurricane damage in recent years, so there are extensive and beautiful coral formations all the way to the border with Belize. In some places there are still fields of elkhorn and staghorn corals, but thin leaf lettuce coral tends to dominate in the shallowest water. The rare honeycomb plate coral has been found on steeper slopes, a smooth sheet-like coral – darkish brown but with white centers to the small polyps. King crabs and spiny lobsters are abundant, and spotted lobsters, marked with masses of white, snowflake-like spots, can also be found.

These reefs are busy with many different fish. Large angelfish are frequently seen, and as well as the more typical butterflyfish there are quite large numbers of the diminutive, and shy, longsnout butterflyfish. Spotted and smooth trunkfish are also common. Sand divers often lie in the sand, but can be spotted when they rise up and dart to another place, before settling back into near invisibility, half buried. They are voracious eaters of other fish, taking prey almost as large as themselves. When a victim passes by, they pounce like a flash, and scores of tiny pointed teeth in their mouths make escape almost impossible.

This southern coastline, now known as the Costa Maya, is about to change dramatically. The sleepy fishing village of Majahual now has a pier for cruise ships and is receiving 3 000 visitors per day, while all along the coast the waterfront has been divided into lots and is being bought up by mostly foreign investors. It is difficult to predict the impact this will have on the wonderful reef world just offshore.

Banco Chinchorro

Surrounded by deep water, Banco Chinchorro is the largest of a series of four atolls in this region (the other three being in Belize). Fishermen have used it for many years, and there are three important lobster and conch fishing cooperatives.

Within the long atoll lagoon there are several islands and a mangrove complex which is home to a healthy population of American crocodiles. There are some beautiful patch reefs scattered through the lagoon, with large boulder star corals, numerous gorgonians, and abundant sponges. Brown and blue chromis shimmer in the bright waters, backed by the brilliant turquoise sand. Dense groups of mutton snapper and schoolmaster rest over these reefs, and there are also abundant red hind, graysbys, coneys, and tiger and Nassau groupers.

The leeward side of the atoll slopes gently down to about 20 meters before a much steeper drop-off. The windward side is littered with shipwrecks, some still visible poking above the water. Below water there are deeply gouged spur and groove formations, the high ridges decorated with blade fire coral and thin leaf lettuce coral in shallower water. In deeper water, macroalgae and soft corals tend to be most abundant. The quite rare fused staghorn coral is also to be found.

Queen triggerfish frequent the outer reef slopes – seen close up they are stunning, with fine black lines radiating from their eyes and two elongated lines of electric blue running back from their mouths in a long mustache. Like all triggerfish they have long, strong teeth – they can crunch up snail shells, but especially favor sea urchins which they flip upside down to get at the less well protected underside. Groups of large tarpon often swim over the reefs, and the deeper slopes are also great places to see larger parrotfish, such as the strangely bump-headed blue parrotfish and the deep blue midnight parrotfish.

BELIZE

Temperature	Belize City: 21-26°C (Dec-Feb), 26-30°C (May-Jul)
Rainfall	1 800 mm; much drier on the reef islands and atolls, while reaching more than 4 meters in the south; wettest from Jun-Oct
Land area	22 169 km²
Sea area	31 000 km²
N° of islands	More than 1 000
The reefs	1 330 km². A most remarkable array of healthy reefs, including patch reefs, a long barrier reef, and three coral atolls. Widely studied by scientists and regarded as some of the best developed reefs in the Caribbean.
Tourism	196 000 visitors, plus 48 100 cruise ship passengers. There are 28 dive centers. Most tourism remains relatively small scale, and the coral reefs are a major attraction.
Conservation	In many ways Belize is ahead of the game with a broad array of parks, some of which at least are now proving highly effective in protecting the reefs and securing the livelihoods of the Belizeans.

For centuries following the arrival of Europeans, the coral reefs off the coastline of Belize must have made this one of the most forbidding shores of the Americas. Far offshore, beyond the continental shelf, were three large atolls, like sentinels, ready to claim unwary navigators. Worse was to come as vessels approached the mainland. Here the shelf edge is traced by a near continuous string of corals coming right up to the surface, often pounded by waves. The few vessels which found a way through the channels in this reef then faced a lagoon studded with shallow reefs.

Charles Darwin was one of the first to write enthusiastically about these reefs, but there has been a surge of interest over the last 50 years. These ancient shipping hazards represent three fabulous coral atolls and one of the best developed barrier reefs in the Caribbean: they have become the most highly regarded coral reefs in the region.

Belize itself is a tiny country, with a small population, a mixture of Mayan, African, and European descent. There is some agriculture, but much of the land remains wild and forested, with large national parks offering excellent opportunities to get close to howler monkeys and toucans. Tapirs and jaguars are present in healthy numbers, but rarely seen. The coastline has extensive mangrove forests, with numerous rivers and streams bringing freshwater and sediments into the lagoon. There are no real fringing reefs, other than a few corals in waters to the south of Placencia, but further offshore the waters

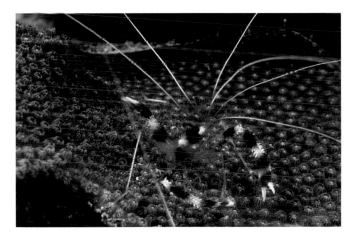

The banded coral shrimp is unmistakable, sometimes known as the barber pole shrimp for its red and white banding. It waves its pincers and long antennae to draw attention to its cleaning services.

undergo a transformation – the sediments settle, seagrasses smother the lagoon floor, and suddenly reefs and islands start to rise up, bright, vibrant, and alive.

Since independence from Britain in 1981, tourism has become the dominant sector of the economy, but it remains mostly small scale, with

Belize

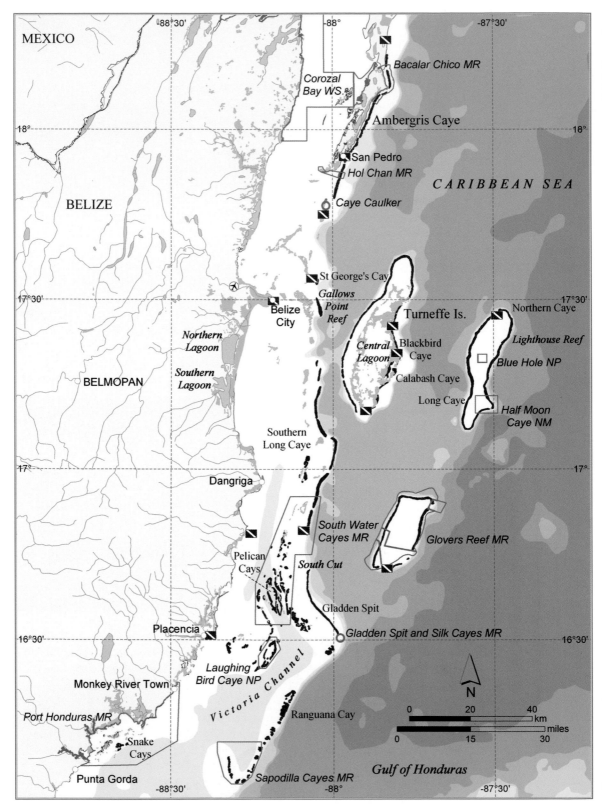

many locally run hotels scattered across the towns and villages, especially along the coast and on the offshore islands. Diving and snorkeling are among the most popular activities, but there are also a number of recreational fishers, and many come to fly-fish for tarpon, permit, and bonefish (see page 124).

Belize has long recognized the importance of its natural resources, and particularly its marine resources, and many countries across the Caribbean would benefit from taking Belize as an example. The Government's Fisheries Department has a Coastal Zone Management Program, which looks at all activities that might impact on Belize waters. This is not easy as it means trying to get agreement from all sectors, including tourism development, forestry, agriculture, business, and conservation groups. Large sections of the country's reefs are now included in marine parks and reserves, and the management of these is closely coordinated with, or even run by, local partners. Recognition of the global importance of these reefs was given in 1996 when these parks were awarded World Heritage status as areas of outstanding importance to the whole world.

The interest in these reefs is truly international. Belize lies at the heart of the Mesoamerican Reef system and it is actively coordinating its conservation efforts with partners in neighboring countries. Scientists from around the world, together with those from Belize's National Coral Reef Monitoring Program, are conducting intensive studies on reef life and health.

There are still localized problems of overfishing, and some of the marine parks receive relatively little active management. The impact of growing levels of sediments and nutrients coming off the land from banana and citrus plantations is also causing some concern.

Some of the earliest studies of the impact of hurricanes on reefs took place in Belize following Hurricane Hattie in 1961, when winds gusted to 320 kilometers per hour and storm surges covered the islands. The waves pounded the reefs, smashing the corals down to bare plains which took some decades to recover.

GATHERING OF THE GIANTS

Prominent "headlands" are prime places for fish to gather when they spawn – the constant flushing by currents carries their eggs far off into the plankton and away from the millions of devouring tentacles and mouths of the reef. It is not such a surprise, then, to learn that Gladden Spit, in the southern barrier reef, is a prime spawning ground. During the full moons of March, April, May, and sometimes June, mutton, dog, and cubera snapper gather here in staggering numbers. Densely packed, the effect is of boiling storm clouds rising up from the sea floor. Spawning takes place just before sunset, with small groups breaking out in frenzied bursts to release milky clouds of eggs and sperm.

These events are a critical part of the great restocking of the reefs – the billions of eggs become millions of larvae, and perhaps thousands survive to adulthood, settling out across the reefs of Belize and probably into neighboring countries. Perhaps because of its remoteness, the spawning aggregation of Gladden Spit has survived the ravages of overfishing, but one fisher has long been aware of this location, a fisher which comes not for the adults, but for their eggs, the caviar of the reef.

For many years the community from the small village of Placencia has known about the seasonal arrival of a whale shark (as pictured below), which they called Sapodilla Tom, just off Gladden Spit. Only in 1998 did scientists first begin to investigate and discovered what must be one of the most spectacular sights of the reef. Whale sharks, the world's largest fish, are usually solitary, but here they found a gathering of these giants, coming to feed on the spawn. Up to 14 whale sharks have been seen in the water at one time, and as the spawning begins they move through, their huge mouths agape, to hoover up the milky bonanza. Research has shown they are mostly young males – through the year they travel as far afield as Mexico and Honduras, but every spawning season, right on schedule, they return to Gladden Spit.

The site was declared a marine reserve in 2000 and today experienced divers and snorkelers can be taken to witness this spectacle.

There are fears that too many visitors will begin to disturb the natural behavior of the sharks, but a series of guidelines has been produced. Tour guides must be trained and have a special license. Visitors are not allowed in between dusk and dawn. They must not touch or chase the sharks, and should try not to block their passage. As interest grows it will be necessary to restrict the number of visitors, but if sound management continues this may remain one of the most magnificent spectacles in the world of Caribbean coral reefs.

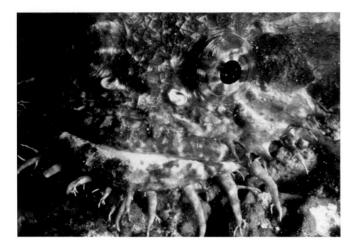

Restricted to the mainland waters from Yucatán to Panama, the large-eye toadfish belongs to an extraordinary family of fish, with several other species also only found in these waters. With powerful jaws they feed, at night, mostly on crustaceans and mollusks.

More recently the barrier reef was badly affected by coral bleaching, both in 1995 and again in 1998, a year which also saw Hurricane Mitch sweep through. These events have added to the impact of losses caused by coral diseases – an impact felt across the Caribbean – and so in most places the corals have changed. Nevertheless, life is still abundant and thriving, and there is every reason to believe that the reefs will remain some of the most spectacular in the Caribbean for many years to come.

There are tight interactions between reefs and mangroves in Belize, and some species, such as the giant rainbow parrotfish, disappear from the reefs if the mangroves onshore are cut down.

TOUR OF THE REEFS
The barrier reef

The Belize Barrier Reef is some 230 kilometers long, not the longest in the Caribbean but certainly one of the most impressive, healthy, and best protected. Moving east across the lagoon it is possible to describe a typical cross-section of the reef. The lagoon itself is 20-40 kilometers wide – seagrasses are common over large areas, but in the north the waters are too murky to support corals. In the south the lagoon is deeper, dropping to more than 30 meters in many areas, and here there are some fascinating reefs. The barrier reef itself rises up from the lagoon on the edge of the continental shelf. There is a wide reef flat, dominated by rock and rubble, and all along this there are numerous small coral cays, crowded with mangroves and, on the higher parts, thatched with other trees and coconut palms.

Beyond the islands waves crash on to the shallow reef crest, and these same waves have carved deep spur and groove formations into the reef slope below. This slope descends, gently or steeply, to a breakpoint, typically at around 30 meters, where the sea floor falls away in a sheer wall. The wall may drop for 130 meters without a break, but steep slopes continue much beyond this, and in the south into waters almost a kilometer deep.

The northern barrier reef

The town of San Pedro, on Ambergris Caye, is the largest tourist center in the country, and a setting-off point for many divers. Here the barrier reef comes quite close to the shore, but is still separated by a lagoon, often about 1 kilometer wide. To the north fishing pressure is low, and the Bacalar Chico Marine Reserve holds some large fish, including Nassau and black groupers, and even jewfish, with eagle and manta rays offshore. Here the reef slope has a complex landscape with sand plains and high coral stacks. Lobsters nestle under the overhangs. Behind the barrier reef there are some very pretty patch reefs, many dominated by huge formations of boulder star corals set against the vibrant colors of the turquoise sand. Look out for bivalves in the crevices between the corals – the flame scallop is a stunning creature with a bright red body, and long tentacles lining its gaping mouth.

The Hol Chan Marine Reserve is one of the great Caribbean success stories. As fishers have agreed to stop fishing in these waters so the numbers of fish have boomed. In a wonderful landscape with deep spurs and grooves there are vast aggregations of grunts and schoolmasters, forming impenetrable masses and often completely obscuring the reef beyond. Cubera snapper are common, and Nassau and black groupers are now unafraid of divers. In the center of the reserve is a deep channel or cut, and there are other deep canyons and caves around. Stony corals include boulder star, lettuce, finger, and various brain corals, while soft corals and rope sponges are abundant. Stoplight parrotfish, blue chromis, and the occasional porkfish all lend flashes of color, while small groups of Atlantic spadefish appear silver against the deep ocean blue. There are still some thriving patches of elkhorn and staghorn on the high ridges.

South of Hol Chan, Caye Caulker is a large, and popular, island, and its offshore waters are now also a marine reserve. The barrier reef from here down to Gallows Point Reef is intermittent, but it rises up to the surface in a few places. Parrotfish, butterflyfish, and surgeonfish are abundant, and the bizarre forms of trunkfish and cowfish are regularly seen. Porgies often venture out over the sand in search of mollusks and crustaceans to eat. On the deeper parts of the reef, below 20 meters, the small yellowcheek wrasse is sometimes found, a striking fish with a purplish to light blue body and a vivid yellow crown from its mouth to the middle of its back.

Central and southern reefs

The barrier reef runs almost unbroken for more than 90 kilometers from Gallows Point Reef to Gladden Spit. These waters are less exposed to heavy swells, being partly protected by the offshore atolls, and the waters are crystal clear.

The shallowest parts just below the reef crest have very deep grooves, and the high buttresses of the spurs are dominated by thin leaf lettuce coral, although other stony corals, including staghorn, finger, massive starlet, and mustard hill, are widespread. Dog snapper and barracuda come even into the shallow waters, and the

Dense schools of dog snapper gather at Gladden Spit to spawn, linked to the full moons of March, April, and May.

constantly shifting and darting forms of chromis provide a busy backdrop. Deeper, there are sea plumes and corky sea fingers, as well as boulder star corals. Tarpon, sometimes in schools of up to 200 individuals, may swim out over the reef crest and down the reef slope. These huge and primitive fish are among the most fecund creatures on the reef: there is a record of a single female (2 meters long) containing more than 12 million eggs.

In a long section around the South Water Cayes Marine Reserve there is a "double barrier" – an inner reef slope drops down into a sandy trough, around 30 meters deep in parts, before rising up again to depths of 12-14 meters. Beyond this ridge, the outer reef slope drops precipitously. Barracuda and schools of Bermuda chub fly across the open space of the wide trench, and eagle rays sometimes pass by. Southern stingrays are frequently seen resting on the bottom. The outer ridge is home to a wealth of corals and sponges, while up close the minutiae of life on the reef continues – arrow crabs and

Deepwater sea fans are widespread on many of the deeper atoll walls off Belize, spread out across the flow of the currents to filter the water for food.

brittle stars often take refuge among the sponges. A great diversity of hamlets – masked, butter, indigo, barred, shy, yellowbelly, black – loiter around the reef. Cleaner shrimps are common – the bright arms and bodies of the banded cleaner shrimp wave for attention, while the nearly translucent forms of Pederson cleaner shrimps have a more subtle beauty, and are often seen lingering near the tentacles of an anemone.

Inshore from the barrier reef, the southern lagoon is host to a maze of shallow lagoon reefs. The waters here are clear and drop to 30 meters or more; the reefs, undamaged by hurricanes and still remote from any human impact, are beautiful. There is a higher abundance of fish on these reefs than almost anywhere else – damselfish, parrotfish, and smaller groupers, plus massive schools of white grunts and others. The elkhorn and staghorn corals which once dominated shallow reefs across the Caribbean are now largely depleted, but in Belize they have not been replaced by algae. There are numerous grazing fish, and plentiful reef urchins, so when these corals died the space was cleared for a proliferation of other corals, notably the thin leaf lettuce coral.

Some of the larger patch reefs in the lagoon, such as those around Laughing Bird Caye, are called faros – they are oval shaped and resemble miniature coral atolls, dropping off at the edges down into quite deep water.

The Atolls
Turneffe Islands

Turneffe is the largest of Belize's three atolls, and quite different from the other two. It is vast, its interior a tangled mat of islands and semi-submerged mangrove forests. There are no deep waters and only a few scattered corals, but seagrasses abound, and these shallow waters are home to substantial populations of crocodiles and manatees. There is an important research station on Calabash Caye.

The leeward, western edge of the atoll has a wide and gently shelving reef slope. In places there are stacks and pillars of coral rock, creating a beautiful landscape. Brain, boulder, and sheet corals grow on the stacks, with soft corals and large barrel sponges widespread in the surrounding landscape. In contrast the eastern shores of the atoll are more exposed and have a similar structure to the barrier reef, with spur and groove formations, and occasional breaks in the landscape, with deep canyons, high buttresses, and sandy plains.

Deepwater sea fans are common on the deeper walls. Black durgon often gather in the waters above the reefs, and permit sometimes swim over the reefs from the lagoon. On the sea floor the strange, cryptic whitespotted toadfish – which is found only in Belize and Honduras – is occasionally to be seen. The rare marbled grouper, too, can sometimes be observed in these waters. An olive brown species marked with small black spots and larger white blotches, this grouper is extremely shy and will retreat into caves when approached.

Lighthouse Reef

Far offshore, Lighthouse Reef is famous for its Blue Hole, filmed by Jacques Cousteau in 1971. Formed by erosion when the atoll was a high island (see page 67 for an explanation of blue holes), it presents a stunning circle of deep Prussian blue against the exquisite turquoise of the lagoon. Some 318 meters wide and 125 meters deep, the blue hole is rimmed by living coral, but deeper down its vertical wall is somewhat devoid of life. At around 30 meters deep there are undercuts and caves, and large stalactites stand testament to the fact that these were once caves filled with air and the constant dripping of freshwater.

The reef falls away on all sides of the atoll with spur and groove formations, becoming a vertical wall in depths as shallow as 10 meters. Sandy patches in the grooves and on deeper ledges are home to large colonies of garden eels, and the occasional sand diver. On the reefs are some huge barrel sponges, and fish are plentiful. Large numbers of ocean surgeon can be seen in the open water, and manta rays sometimes pass by, but the small life on the reef itself should not be overlooked. Arrow blennies are common in some recesses on the wall – these slender, red and yellow fish, with pointed snouts, are quite unlike other blennies. They rarely sit in holes and are usually seen drifting about with their tails twisted hard round in a U-shape. This acts like a cocked gun and, despite the apparent drifting, these tiny fish are always ready to pounce. When they come within striking distance of even smaller fish, such as masked gobies, they release the full power of their taut tails and are propelled forwards at lightning speed to make their catch.

Half Moon Caye is a tiny coral island on Lighthouse Reef which has become a key nesting ground for 4 000 red-footed boobies, as well as the predatory magnificent frigate birds. Hawksbill and loggerhead turtles nest on the beaches, and the island is strictly protected.

Glovers Reef

Glovers Reef is a beautiful atoll – its wide lagoon is 6-18 meters deep, but is bejeweled with more than 700 patch reefs, all decked in healthy corals and shimmering fish. The clarity of the surrounding waters is renowned, and there is still sufficient light for stony corals to grow in depths of more than 100 meters on the reef slopes. The western slopes present a highly sculptured reef surface – there are wide patches of elkhorn in some places, but other stony corals include yellow pencil, finger, thin leaf lettuce, large cactus, and a variety of brain corals. Off the eastern shore the reef slope again becomes a vertical wall, even overhanging in parts, and covered with tube, rope, and encrusting sponges. Not far offshore the waters are more than a kilometer deep. In places there are narrow ledges, and elsewhere there are small caves and recesses, often guarded by Nassau, black, tiger, and marbled groupers.

Bull sharks and blacktip sharks may be seen on all the Belize reefs, but especially in more remote locations. They are quite common off the northeast point of Glovers Reef (Shark Point), where the reef is exposed to heavy swell, and where they are sometimes joined by tiger and hammerhead sharks.

All of Glovers Reef is protected, with a closed wilderness area, a large conservation zone (where diving, but no fishing, is allowed), and general use zones. A research station on Middle Cay is providing critical information about the nature and health of the reefs. Although the eastern reefs were damaged by Hurricane Mitch, the patch reefs and those on the western shore were protected from the large waves.

The southern end of the Belize Barrier Reef around Gladden Spit and Glovers Reef offshore, seen from the space shuttle. The elliptical shapes of the faro reef formations can be seen in the lagoon.

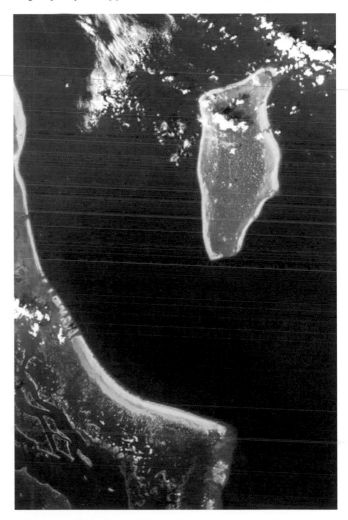

Honduras

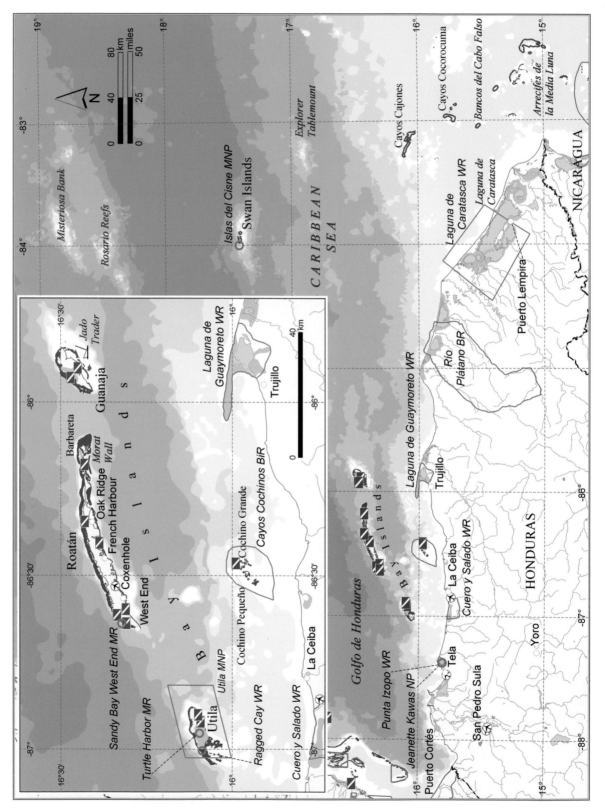

HONDURAS

Temperature	La Ceiba: 20-26°C (Jan), 25-31°C (Jun)
Rainfall	2 850 mm; peak rainfall Sep-Feb
Land area	112 851 km^2
Sea area	238 000 km^2
N° of islands	More than 100
The reefs	810 km^2. All the reefs are offshore. The reefs around the Bay Islands are spectacular, with deep walls and a high diversity of species.
Tourism	408 000 visitors. There are 33 dive centers, all on the Bay Islands and Cayos Cochinos, and indeed diving is one of the major sectors of the economy in these islands.
Conservation	There is a strong awareness among both the local people and the government of the importance of healthy reefs, and although more protected areas are needed, the existing levels of protection are already valuable.

Mainland Honduras is not a place for coral reefs – the coastline, although remarkably beautiful and rich in wildlife, is deluged by rivers, thick with sediments. Not far from shore, however, the waters deepen and become clear (Columbus first used the name Honduras, meaning "depths", because of the deep waters all around). Scattered through these deeper, cobalt blue waters are a large number of islands and fantastic reefs, shimmering with life.

Human settlements in Honduras go back millennia: the magnificent Mayan settlements at Copán date back to 1000 BC, while the Paya people probably inhabited the Bay Islands from around AD 400. In the east, the Miskito people make a living through agriculture, fishing, and hunting. The Bay Islands were, for many years, a base for large numbers of English and Dutch pirates who harried the Spanish ships as they left this coastline loaded with treasure. These islands, and the eastern Mosquito Coast, later became a British protectorate. Control was relinquished to the Spanish in the mid 19th century, but even today English remains widely spoken.

Along the mainland coast there are still some wild areas, including the Río Plátano, a vast biosphere reserve whose international impor-

tance has been recognized through its designation as a World Heritage site. There are extensive areas of mangrove forests, lining coastal lagoons and tracing the banks of rivers far inland. These are home to crocodiles, caymans, and manatees, while large fish such as sharks, tarpon, snook, and kingfish come into these areas from the ocean to breed or to hunt. Turtles nest on many of the adjacent beaches.

Adult blue tangs form large schools which travel over the reef and occasionally dive down to graze, overwhelming the territorial claims and defensive efforts of damselfish.

The best coral reefs are formed around the Bay Islands which lie along the Bonacca Ridge, and around the Cayos Cochinos to the south. There are also scattered reefs on the wide shallow shelf to the east of the country. Both shrimp trawling and lobster fishing are major industries on this shelf, but remarkably little is known about the coral reefs themselves. Honduras owns the Swan Islands, far out to the north, and the Rosario Reefs and Misteriosa Bank on the Cayman Ridge also fall within its territory.

Hurricane Mitch, which swept across Nicaragua, Honduras, and Belize in 1998, was a tremendous storm (category 5) which brought winds of 250 kilometers per hour and vast amounts of rain. For 48 hours it stalled over the Bay Islands. On land it wrought enormous damage to trees and buildings, but remarkably the coral reefs were only partly affected, and still

The silky shark is an oceanic species, occasionally venturing over deep reefs. "Bluewater" diving, where divers wait in the ocean depths to see what swims past, can reveal a range of large predators.

remain in very good shape. In 2001 another hurricane, Iris, again left the reefs largely unscathed.

There is widespread interest in conservation in Honduras. Officially, all of the waters around the Bay Islands were declared protected in 1997 although there have been few changes to management. More active protection is provided in the Sandy Bay West End Marine Reserve on Roatán, which is largely financed by the nearby hotels. The waters around the Cayos Cochinos are also now receiving quite effective levels of protection.

TOUR OF THE REEFS
Bay Islands

Spectacular reefs surround the Bay Islands, and these are home to a remarkable diversity of life. Some 309 different reef fish have been observed in the waters around West End on Roatán alone.

All of the islands rise up from deep water. In places there are bright sandy bays, but elsewhere ironstone – sharply eroded limestone – forms ragged cliffs. In the nearshore waters reef life often begins quite close to land with a gently sloping plain, and there are excellent areas for snorkeling. Fire corals are abundant, with pencil corals and boulder star corals amongst wider tracts of sea rods and sea plumes. In one or two places, notably off the east of Guanaja, staghorn and elkhorn still form wide forests. Damselfish are always abundant in these shallow waters, and juvenile fish – grunts, snappers, young butterfly-fish, and angelfish – often move out to these areas from the protection of mangroves and seagrasses, before moving on to the deeper reefs.

Typically, the shallow reef ends at depths of 10 to 15 meters, plunging down to depths of 50-80 meters. In places these walls are rent apart by spectacular canyons, and are deeply incised by caverns and tunnels. Towers rise up from the deeper water. Some of these structures have been built by the growth of corals; others have been influenced by volcanic activity. In quite a few places there are also signs of erosion from the ice ages, when sea levels were much lower, and when seabirds soared over the now submerged cliffs.

The steep walls are smothered in life – with sea fans and sheet corals, as well as a profusion of sponges. In addition to rope sponges and tall tube sponges, there are some very large barrel sponges, including the netted barrel sponge which can reach 1.5 meters in height and has a brownish yellow or olive green exterior traced with fine ridges. In deeper areas and on the steepest walls there are abundant black corals – including bushy, feathery, and fan-like forms. Cryptic fish are common, while scorpionfish, toadfish, and even seahorses are regularly seen by more observant divers, or at least by those who have learnt the knack of how to spot them. Swirling masses of silversides often collect amongst the most complex sculptured towers and hollows, and eagle rays sometimes soar over.

Quite a variety of mollusks may be found on these reefs, including the largest Caribbean snail, the horse conch – recorded to 48 centimeters in length, they are not conch at all, but a type of tulip shell. They are powerful predators which catch other snails and are strong enough to prevent them retreating into their shells for protection. Holding them thus, they slowly begin to eat them alive. At the other end of the spectrum the Bay Islands are host to a wealth of shell-less nudibranchs, such as the gold crowned and the purple spotted sea goddesses. Although only 1-3 centimeters long, these are regarded by many visitors as the most beautiful animals on the coral reef.

Peering into caves and caverns can reveal new creatures. Orange cup corals often cover the walls; their bright colors, seen with a flashlight, are quite breathtaking. A little less common is the exquisite, but fragile, rose lace coral – closely related to fire corals, these corals have no algae in their tissues and so do not need to be in the sunlight.

The larger hollows and caverns are often home to schools of glassy sweepers, waiting until nightfall to go out and forage. Smaller spaces give shelter to cardinalfishes. These fish are common on all Caribbean reefs, but their diminutive size and skulking habits mean that they are often overlooked. There are 22 different species found across the Caribbean and many share an extraordinary breeding habit – once the eggs are fertilized the males protect them from predators by holding them in their mouths until they hatch. This habit, known as mouth-brooding, is also undertaken by the yellowhead jawfish. Brooding males can sometimes be seen with swollen mouths, slightly agape, with a huge ball of eggs inside.

Out in the open water the theater of life is more unpredictable, but there is almost constant activity near many reef walls. These waters are host to almost all of the big silvery predators – jacks; barracuda and their smaller relatives, the southern sennet; and also the relatives of the tuna, including bonito and cero. Some divers make "bluewater" dives in these waters – they are dropped in the open ocean where the water may be hundreds of meters deep and simply wait to

During the ice ages lower sea levels meant that the walls around the Bay Islands were once sea cliffs. Many of the present-day caves were probably carved out at this time by waves at the sea surface.

see what swims past. This may include various sharks (typically ocean-going species such as silky sharks), but also bonitos, manta rays, marlin, or sailfish. Such diving is usually only for more experienced divers – with no sea floor as a reference it is easy to lose all sense of perspective and divers must be careful not to sink too deep. Off the northeast coast of Utila whale sharks are quite regularly seen swimming near the surface.

Over sandy patches both on the shallow reef slopes and in breaks in the reef walls there is a great deal to be found. In the distance may be seen wide fields of garden eels, rising like fine fronds and gently sinking down to disappear in the sand as a diver approaches. Flounders are common, and groups of feeding goatfish may drift over the sand. Where there is more rubble,

THE CORAL REEFS OF GUATEMALA

Sandwiched between the Bay Islands and the Belize Barrier Reef, the short Guatemalan coastline tends to be overlooked by coral reef scientists as well as by divers.

This is a coast with high rainfall, heavy sediment loads, and muddy sand offshore, not the place for thriving coral reefs. Remarkably, there are a few coral reefs growing further offshore. The dominant species are generally those more tolerant of the constant rain of sediments, such as massive starlet corals. These are also important waters for turtles, with hawksbill, green, leatherback, and loggerhead all coming to nest in the Punta de Manabique Wildlife Refuge. Manatees are often seen in some of the bays.

Yellowhead jawfish build themselves complex homes consisting of small chambers in the rubble, connected by a tunnel to the surface. They retreat backwards into them if threatened.

yellowhead jawfish are quite common. These small, pearly fish, with bluish fins and a yellow tint to their heads, live in small colonies and build burrows. During the day they hover in the water above their burrows, picking at tiny food particles in the plankton, but they are wary, and as divers approach they drift downwards and will sink, tail first, into their holes if worried. Within just a few minutes they re-emerge, and can get quite used to the presence of patient observers.

Turtles are relatively common – they are seen on almost every dive around the West End of Roatán, and considerable numbers of both loggerhead and green turtles nest in the Turtle Harbor Marine Reserve on Utila.

A few wrecks have been sunk in these waters, the largest of which is probably the *Jado Trader*. Sunk off Guanaja in 1987, this 70 meter boat is now smothered in sponges and soft corals, and larger snappers and jacks drift in schools over its large hulk.

In the more exposed places, strong currents offer experienced divers opportunities for drift diving. One of the best spots is the Morat or Barbareta Wall which lies quite far offshore to the south of eastern Roatán. It is not only divers who use these currents however: many fish congregate on current-swept headlands to release their eggs. Perhaps the best known spawning site in the Bay Islands is off the far eastern tip of Guanaja, a site known as Caldera del Diablo (Devil's Cauldron). Here, during the periods around the full moons of January and February, big groupers, including Nassau groupers, black and yellowfin, gather in huge numbers to spawn.

Cayos Cochinos

Only 16 kilometers from the mainland, the Cayos Cochinos are a group of 13 tiny coral cays and two larger, hilly islands. They are surrounded by coral reefs, teeming with life, which are protected in a biological reserve where no anchoring or commercial fishing is permitted. A few of the cays are important nesting grounds for seabirds such as the royal tern and the brown pelican, while the larger islands are also home to a large, pink boa constrictor which is unique to these islands.

In some ways the reefs are similar to the Bay Islands, with some steep walls to the north, but they also have some stunning areas of shallow water which make for excellent snorkeling. There are several species which are rarely seen in the Bay Islands, including batfish, frogfish, and the small quillfin blenny. Slate pencil urchins are seen in the shallows and there are many colonies of bright blue bell tunicates. Snorkelers often find themselves being watched by reef squid, although these wary creatures rarely allow a close approach.

In some places there are complex seascapes, with towers and ridges rising up from an undulating base – shimmering schools of silversides and densely packed masses of grunts drift in these areas, lending a magical appearance.

Far out into the Caribbean Sea, two high islands and one tiny coral cay make up the Swan Islands. Red-footed and brown boobies nest on the islands, which may also be important for nesting turtles, but little information is available. The same distance to the north again there are coral reefs, but no islands, rising up on the Cayman Ridge – the Rosario Reefs and the Misteriosa Bank. All of these areas have highly productive fishing grounds for conch, lobster, and various fish.

NICARAGUA

Temperature	Puerto Cabezas: 23-27°C (Dec-Feb), 26-29°C (May)
Rainfall	2 970 mm, rising to 4 430 mm in Bluefields and to more than 6 meters near the border with Costa Rica; peak rainfall Jun-Dec
Land area	129 047 km²
Sea area	127 000 km²
N° of islands	More than 100
The reefs	710 km². There are numerous reefs which remain little known, but are undoubtedly influenced by the high levels of nutrients and sediments in the water. Many are well adapted to this.
Tourism	486 000 visitors. Most go to the central and western parts of the country. There are four dive centers, but only one of these is on the Caribbean coast.
Conservation	There is a considerable interest in ensuring that development across Nicaragua is sustainable but the Caribbean coastline has received little attention to date.

Nicaragua's wild Mosquito Coast is a largely unknown territory in the world of coral reefs. Islands and reefs lie scattered across one of the widest continental shelves in the Caribbean, but these have had little attention and remain difficult to access. Inland, the wide coastal plains remain largely undeveloped. The area fell under British protection from the late 17th century, with a "king" appointed from the Miskito people – it became fully incorporated into Nicaragua only in 1893, and English is still the main language in some areas. Although the Amerindian peoples (mostly Miskito, but also Sumos and Ramas) were devastated by slavery and European diseases, many remain, forming the majority of the population away from the towns, and living from a mix of fishing, hunting, and agriculture.

The soils are very poor, and so only pines and savanna survive in the north, with rainforest further south. Numerous rivers carry 90 percent of the country's rainfall into the Caribbean. Manatees are still widespread in the mangrove-lined river mouths, and sharks, which can be highly tolerant of freshwater, regularly enter the rivers. For many years a population of sharks in the large Lago de Nicaragua were thought to be a unique, freshwater species, but in fact they are

The powerful bull shark is able to survive in freshwater and regularly migrates up the rivers on the Caribbean coast of Nicaragua.

Nicaragua

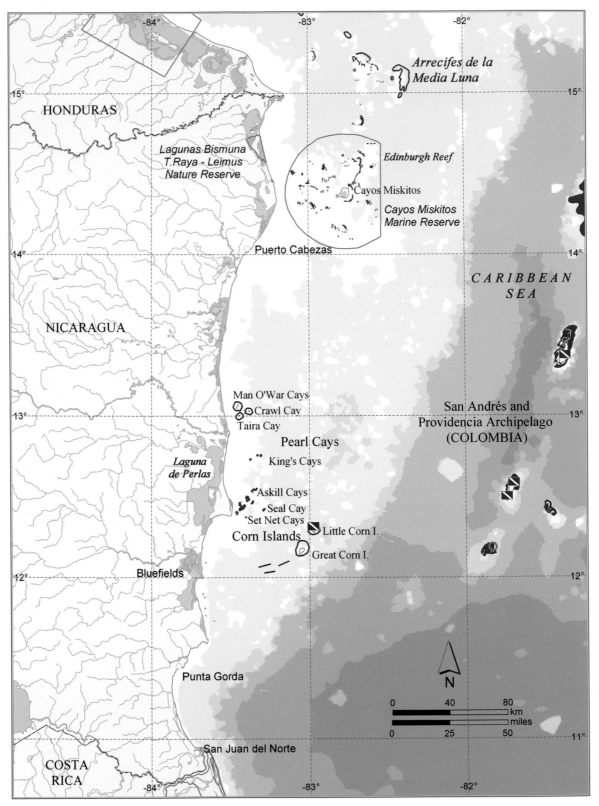

bull sharks which migrate up the Río San Juan from the Caribbean.

Instability and poverty wracked this country for many decades until 1990, and while today there is a secure peace and a sense of optimism, the Caribbean coast remains poor and undeveloped. Parts of the coast in the north are still quite dangerous for travelers.

Offshore the waters drop to depths of 20-40 meters, but then continue at this depth across much of the continental shelf. This shelf is fed by nutrients from the many rivers, and from deep water upwelling to the east, and so the waters here are highly productive, supporting important lobster (see page 108) and shrimp fisheries. Large parts of the continental shelf are covered in seagrass beds, which may be the most extensive in the Caribbean and are the main feeding ground for many of the green turtles that nest in Costa Rica (see page 111).

From the small amounts of available information it seems that the coral reefs along this coastline are not as healthy as might be hoped for such a wild area. The country was badly affected by Hurricane Mitch in 1998, and had been even more badly affected by Hurricane Joan in 1988. The latter produced waves up to 15 meters in height, destroying many corals, while smothering others in sand and rubble. Joan also damaged vast areas of forest inland, which created high levels of erosion and brought even more sediments to the coastal waters.

Direct impacts are felt from the prawn trawling industry – up to 90 percent of catches are discarded, mostly dead, as bycatch, and this includes large numbers of juvenile reef fish. Turtle fishing is also widespread and largely uncontrolled. These problems should be set against a background of very low human impact in other spheres – some reefs are damaged, but as more work is done in remote areas it is likely that extensive and healthy coral reefs, and important fish populations, will be discovered.

TOUR OF THE REEFS

Scientists divide Nicaragua's coastal shelf into three bands. The inner shelf, which extends about 15 kilometers from land, has few reefs and is strongly influenced by the rivers and sediments which pour into the coast. A coastal boundary current tends to keep the sediments concentrated within this area. The central shelf has clearer water and this is where most of the known coral reefs occur, as described below. Finally, the shelf edge and shelf slope have the clearest water of all. This area remains little known; there are some wide algal banks and some coral formations, but much more remains to be explored.

Large eyes and red coloration are clues to the fact that the longspine squirrelfish is a nocturnal species. Red looks black as the light begins to fade and these fish hunt intently for shrimp and other creatures.

The two Corn Islands are the most regularly visited coastal islands. Great Corn Island has a population of more than 4 000 people on an island only 5 kilometers in length, while Little Corn Island has been partly developed and offers some diving opportunities. There are scattered coral reefs all around both islands, although some areas close to Great Corn Island are degraded, probably due to sewage pollution.

Further from shore, boulder star and sheet corals form rich gardens, along with tall sea plumes, descending down to 20 meters. Angelfish, triggerfish, and parrotfish add color to the reef scenes, and southern stingrays are often observed in the adjacent sandy areas. Sharks are also quite commonly seen – nurse

A TERRIFYING TRADE

The lobster fisheries across the shelf areas of Nicaragua and Honduras are a huge and important trade. Exports of lobster tails to the USA and other wealthy nations bring in $30-40 million per year for each country, and the fishery is a key source of employment. Lobsters are caught in two ways – with traps and by scuba divers using gaffs (long hooks). While trap fishing is relatively safe, the use of scuba equipment brings with it tremendous dangers.

Only a handful of the lobster divers have had any form of training. Most dive with very basic equipment – no depth gauge, pressure gauge, or watch, and tanks that may be 20 years old and never pressure tested. They usually make over ten dives a day, often to depths of more than 30 meters. Without the correct equipment they often run out of air and have to make an emergency ascent. The toll this takes on the population is enormous. There are probably several thousand disabled former divers – suffering weakness and paralysis as a result of decompression-related incidents. Over the last decade more than 120 have died in Honduras alone.

Despite the dangers, becoming a lobster diver is one of the only means of entering the cash economy, particularly for the Miskito people, and so many young people are keen to learn. In Miskito traditions, the Liwa Mairin is the spirit associated with the sea, often represented as a mermaid. She is believed to safeguard the sea's resources and indeed to punish those who take too much from the sea. These beliefs are used to explain the injuries sustained by many fishers. Divers claim to have seen the Liwa – perhaps while suffering from nitrogen narcosis from diving too deep.

Overfishing, illegal fishing by foreign vessels, and the taking of young and gravid (egg-carrying) lobsters are all adding to the problems. As the numbers of lobsters decrease, divers are being asked to dive deeper and more often to collect their quarry, while the profits being made are falling.

Closing this trade would be extremely destructive to the poor communities of the region, but there is an urgent need to stop the dangerous diving, perhaps replacing it by trap fishing. In the meantime some groups are working to provide proper training for the divers, and a few decompression chambers have been installed. The governments need to reduce overfishing by closing off areas, by reducing the numbers of fishers, and by putting a stop to illegal vessels.

Much of the pressure to overfish is driven by ever growing demands in developed countries, so many conservation organizations are recommending that people should boycott Caribbean spiny lobsters, both when visiting these countries and when purchasing shellfish at home. It is not that they want the industry to stop, but a reduction in demand may help to bring the industry back to a more sustainable footing.

sharks rest on the bottom, while blacktip and even hammerhead sharks may pass through. The visibility can be low, particularly in the rainy season, but the rich life still makes this an impressive area. One species which prefers slightly murky waters is the lookdown, an extraordinary jack with a wafer thin, but very deep, body of the brightest silver. They are usually seen in schools, and feed on fish and crustaceans.

The Pearl Cays lie much closer to the mainland. They are uninhabited, but there is a small lobster processing plant, and turtle fishers sometimes stay on the islands. In the 1970s the corals around these islands were described as "luxuriant and well developed", but some areas closest to shore have since been smothered by sediments and have died. Healthy areas remain in the north and east. Boulder brain, mustard hill, and great star corals predominate, but the thin leaf lettuce coral does well in the more silty areas. With its vertical blades it is able to shed large amounts of sediment and avoid being smothered.

Parrotfish and surgeonfish are widespread, grazing on the plentiful algae. Squirrelfish are also abundant, hovering low down or in recesses amongst the corals – the large eyes of these fish and their red coloration are both adaptations to their nocturnal habits. Red is one of the first colors to disappear as the light fades, so when they set out at dusk to feed across the reef and seagrass areas they appear black, and merge into the shadows, while their large eyes enable them to spot crustaceans and other food. Barracuda regularly visit the reefs, and in the deeper areas off Seal Cay nurse sharks and dog snapper are not uncommon.

The most extensive areas of islands and reefs are the Cayos Miskitos. Most of the islands are dominated by mangroves, with numerous patch reefs and large seagrass beds in the waters all around. Despite their importance, remarkably little is known about these reefs. A marine reserve has been declared, but there have been few efforts to enforce any rules to date.

COSTA RICA

Temperature	Puerto Limón: 22-28°C (Dec-Mar), 23-30°C (Sep)
Rainfall	3 390 mm; no distinctive wet or dry season
Land area	51 608 km²
Sea area	566 000 km²
The reefs	Less than 100 km². The Caribbean coastline has only a few reefs, mostly towards the southeast.
Tourism	1 106 000 visitors. The natural beauty of this country is a major attraction. Most tourists come to the central highlands and Pacific coast, but growing numbers visit the turtle nesting areas and the beaches of the southeast. There are 11 dive centers, but only one on the Caribbean coast.
Conservation	Concern for the natural environment has driven a strong conservation ethic in Costa Rica; however some of the coastal reefs in the Caribbean remain unprotected, while runoff of pollution and sediments has adversely affected a few areas.

Much of Costa Rica's Caribbean coast is a long sweep of sand, pounded by surf. Behind this, there are extensive, wet tropical lowlands, and in places these remain as a wilderness of swamps and thick forests. Elsewhere, wide areas have been cleared to make way for cattle grazing and for vast banana plantations. Moving inland, the country rises up to a great spine of volcanic peaks and fertile highlands, densely populated. The Pacific coast presents a stark contrast to the Caribbean, with its rocky twisting shores and offshore islands.

Many travelers come to Costa Rica to enjoy its great natural beauty, and the opportunity to hike on the volcanic slopes inland should not be missed. The warm wet climate and rich soils support slopes that are thickly clad in cloud forests. Strange twisted trees have thick, shaggy coats of mosses and luxuriant bromeliads. Within the volcanic craters there are hot sulfur springs, steam vents, and boiling mud pools. Some are sufficiently active that it is possible to witness (from a safe distance) lava flows and minor explosions of incandescent rocks, a spectacular firework display, especially at night.

The Caribbean coast is not noted for its coral reefs. Along most of its length the surf and shifting sediments have prevented coral growth altogether, but towards the south there are a few places where rocky headlands have provided a foothold for corals. Some of these are

within protected areas, but unfortunately there are still problems of sedimentation and pollution associated with deforestation and the intensive use of agrochemicals.

TOUR OF THE REEFS

The most northerly corals are found near the major port of Puerto Limón. With powerful sea all around, the small reef patches are concentrated on the leeward (western) sides of offshore islets and embayments, but pollution is a growing problem.

A 5 kilometer long reef, Costa Rica's largest, traces the headland around Punta Cahuita. It has a shallow reef crest, which then drops with deep spur and groove formations to a sandy sea floor at 15 meters. Further offshore there are a few shallow banks with scattered corals, but the most abundant corals lie in the lagoon behind the main reef. Here there are a few small but vibrant fringing reefs.

Damselfish are abundant, with sergeant majors parading their bright markings high up in the water, while yellowtail and dusky damsels skulk nearer the sea floor. Bright gangs of stoplight, striped, and yellowtail parrotfish are common, and there are also large, busy schools of surgeonfish – ocean and blue surgeons, and

Costa Rica and Panama

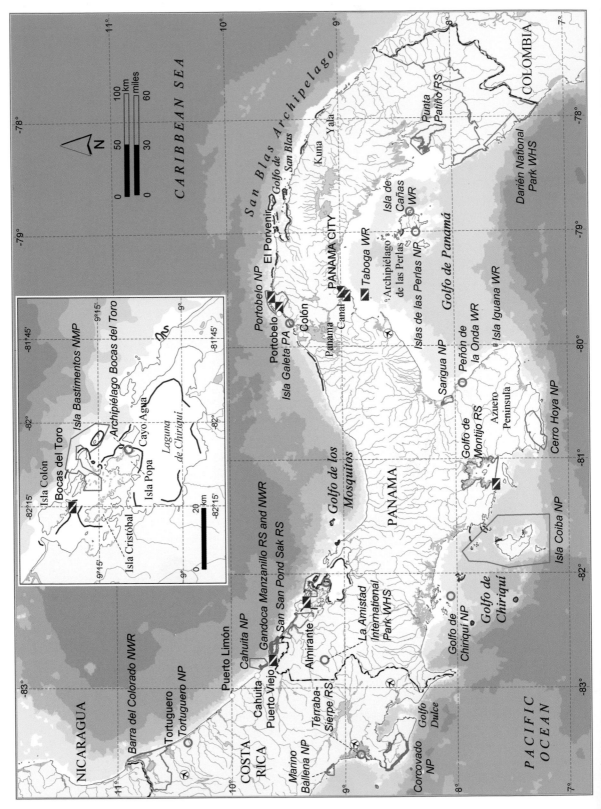

doctorfish. All of these fish are algal grazers, and their abundance is a pointer to two things – firstly, and obviously, there is a wealth of algae, but secondly there are almost no grazing sea urchins. The die-off of the long-spined sea urchin in 1983 was first observed in the waters off Panama, and in these nearby waters they have made little or no recovery, leaving the algae pretty much to the fish.

There are a number of stresses on these reefs, including high levels of sedimentation, but a more unusual impact took place in 1991 when an earthquake lifted up a great section of this coastline. Some areas were raised as much as 1.5 meters, exposing parts of the reefs to the air, and killing the corals.

The final area of coral reefs lies along the short stretch of coastline from Puerto Viejo to the Panama border. Here there are rocky headlands and small beaches, and small fringing reefs drop down in some very pretty buttresses from 2 meters to 6 meters. These reefs are home to a variety of star corals, knobby and symmetrical brain corals, massive starlet corals, and lettuce corals.

The usual denizens of Caribbean reefs abound, with great crowds of grunts waiting out the day in the shallow water, including caesar and Spanish grunts. Barracuda regularly come and patrol the reefs. There are also quite a few banks offshore, rising up from sandy surroundings to within 8 meters of the surface. Their outer slopes are home to more unusual whip corals, wire corals, and sea whips, as well as the sunray lettuce coral, cactus corals, and rough star corals. There are also spectacular barrel sponges reaching to 1.5 meters in diameter, and tall organ pipe sponges.

Many of the reefs fall within the Gandoca-Manzanillo Wildlife Refuge, but this site is famed more for its marine turtles than for coral reefs. Four species regularly come up to nest on its black sand beaches – loggerhead, hawksbill, green, and most importantly one of the largest nesting populations of the giant leatherback turtle in the southern Caribbean. Poaching was having a severe impact on these populations, but since 1986 regular patrolling of the beaches by volunteers has reduced this to a minimum, and these magnificent beasts now appear to be slowly increasing in number.

Young hawksbill turtles setting off on their hazardous journey to the sea.

TURTLE COUNTRY

Through most of the Caribbean, marine turtles are now quite rare, and even the busiest nesting beaches are visited by only a few tens or hundreds of nesting turtles each year. Costa Rica, by contrast, gives us a small taste of an earlier Caribbean, where turtles were once superabundant. Along the northern Caribbean coast, in the Tortuguero National Park, well over 10 000 green turtles come to nest each year. This is more than ten times the number recorded from any other location in the Caribbean.

For many years now, volunteers from around the world have patrolled this coastline at night. Their work has reduced poaching, and has provided invaluable insights into the number and behavior of the turtles. The number of turtles coming up to nest has been slowly increasing over the past two decades, but this coastline still holds natural hazards for these ancient reptiles. Offshore, bull sharks occasionally take smaller turtles, while on land jaguars have been known to come out of the forest to kill and eat the nesting females.

The beauty of this coastline, with its wild beaches, long lagoon channels, and dense forest is attracting ever increasing numbers of tourists – more than 20 000 a year – and their interest in watching the nesting turtles has brought an important shift to the economy. All tourists must be accompanied by a local guide, who is paid by the visitors. The value of the turtles alive is already far greater than would be the case for their bodies or their eggs.

Turtles also provide wonderful spectacles on the more remote beaches of the Pacific coast. One species in particular, the olive ridley turtle, exhibits mass nesting, known as *arrabidas*, on two beaches in the northwest (not on the map). Although some turtles come to nest year-round, there are periods between July and October, during the last quarter of the moon (the darkest nights), when hundreds – perhaps thousands – of turtles arrive each night. This frenzy of nesting means that many turtles actually dig up the eggs of other turtles and, recognizing that many eggs are thus doomed anyway, a controlled harvest of some eggs is currently allowed.

PANAMA

Temperature	Colón: 23-28°C (Nov-Dec), 25-30°C (Apr)
Rainfall	3 460 mm; drier from Jan-Apr
Land area	74 697 km^2
Sea area	332 000 km^2
N° of islands	About 450 on the Caribbean coast
The reefs	570 km^2. There are extensive reefs in the east and the west, with high cover of healthy corals and diverse fish communities. Central areas have more scattered fringing reefs, some of which are now quite degraded.
Tourism	479 000 visitors. There are 16 dive centers, nine of them on the Caribbean coast. Diving is popular in Bocas del Toro and the central coast, but is not possible in San Blas, although these islands offer excellent snorkeling.
Conservation	Many of the reefs have some degree of legal protection within protected areas or areas controlled by Amerindian groups. To date there is little active management, but the reefs have remained relatively healthy.

There is a widely held belief that the word Panama is derived from the name of a coastal village and translates, from the Cueva dialect, as "place of many fish". Whether true or not, the Caribbean coast of Panama (see page 110) lives up to such a name. Fish and other life are abundant along a coastline traced by long lines of islands and coral reefs set against a backdrop of mangroves, rainforest, and high mountains.

Like its South American neighbors, Panama

Modified scales give boxfish, such as this scrawled cowfish, a form of body armor, but mean that they cannot flex their bodies – only their eyes, mouth, fins, and tail are free to move.

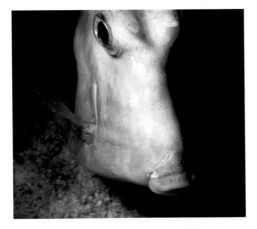

lies sufficiently far south to avoid the ravages of hurricanes, and this has helped in the development of spectacular coral reefs. The most extensive reefs are those of the San Blas Archipelago, homeland of the Kuna people.

Along the central Caribbean coastline there are a few more fringing reefs as well as coral-dominated slopes, while in the west the Archipiélago Bocas del Toro has extensive and beautiful reef formations.

At the time of the Spanish arrival, Panama was already a densely populated agricultural nation. Although the Amerindians were decimated following Spanish settlement, many still remain: the Guamí people are widespread on the western Caribbean coast and islands, while the Kuna live in near autonomy in the east. Both groups have villages along the coast and on the offshore islands, and live in close contact with the sea and the coral reefs.

Panama is the narrowest of the Central American countries, where the Pacific and the Caribbean come within 60 kilometers of one another. On a clear day it is possible to see both sides from various high points across the country, and indeed it is possible for divers to dive in both oceans on the same day.

This close connection led to speculation about the building of a canal as far back as 1524. The first efforts at construction did not take

Stripy sergeant majors and young schoolmasters swim around a pillar coral in bright shallow water.

place until 1880, and were thwarted by harsh conditions, but the canal was eventually opened to shipping in 1914. The Panama Canal today dominates the economy, with 13 000-14 000 vessels passing through each year. There are constant fears about pollution and the risk of pest species being carried from one ocean to the other.

Agriculture is also important across the country, ranging from traditional shifting agriculture to large industrial banana plantations in the west. The latter produce considerable amounts of pollution which threaten some of the coastal reefs.

Many of the coral reefs are protected within national parks and wildlife refuges, while other areas fall within the legally assigned lands of the Amerindian groups. In practice there are limited efforts at management, but there is ongoing research in many areas which will be important in detecting change and in future efforts to protect the reefs more comprehensively.

TOUR OF THE REEFS
The west

Taking a boat through the complex of islands, mangroves, and shallow waters of the Bocas del Toro (the name describes an archipelago, a province, and a town) is a rewarding experience. Far from any towns or roads, the thatched dwellings of the Guamí people are widespread, and their small canoes and sailing boats ply these waters. Dolphins quite often pass by in small groups, and pelicans regularly fly overhead.

On a calm day houndfish sometimes break the surface, fleeing the boat. These are the largest of a group of fish known as needlefish, which appear like a long splinter of silver, with a needle-like mouth. Seeming to burst out of the water, they either jump sleekly or skitter across the surface, for 50 meters or more, powered by just the smallest tip of their tails in the water. There are stories of fishermen walking in shallow water at night being impaled by these fish, blinded as they are by the fishers' lights.

The town of Bocas del Toro has a large number of English-speaking people and in many ways seems more like an eastern Caribbean outpost than part of Latin America. From here locals, visitors, and scientists can travel to any of the six larger islands or hundreds of smaller cays. There are extensive fringing reefs, best formed away from the somewhat rough waters of the north-facing island edges. Even the shallowest waters have extensive corals, including some of the largest stands of elkhorn corals remaining in the Caribbean, as well as areas of staghorn, finger, and lettuce corals.

Foureye and spotfin butterflyfish pick their way through this bright scenery, with larger angelfish making an occasional appearance. Bar jacks are common, swimming past in flashes of silvery blue. Far harder to spot in the shallow areas is the spectacular greenbanded goby, sometimes seen resting under sea urchins or in the shadow of small rocks. These fish have a broad reddish streak through their eyes, and dull green bodies lit up by numerous fine stripes of iridescent lime green.

Over the bright shallow sand, eagle rays sometimes come to forage. Although more often seen doing a majestic fly-past, eagle rays, like their more sedentary cousins the stingrays, may be observed foraging in the sand for food. As they dig, great clouds of sand are thrown up,

often attracting other fish which are looking for scraps.

The deeper reef slopes have large quantities of sponges and algae, but still there are rich coral communities. Although larger fish are not common, hogfish are occasionally seen here. The larger males are truly impressive fish, which can reach a meter in length, with darkened foreheads and elongated dorsal spines; they patrol a territory on the reef which may contain up to 15 females. Banded coral shrimps can usually be seen, resting in the recesses or on the surfaces of sponges. One of the larger cleaner shrimps, they tend to back away when approached, but occasionally can be tempted to venture on to, and clean, a gently proffered hand.

The central coast

The coral reefs along the central parts of Panama's Caribbean coastline are mostly fringing and patch reefs. They are not well developed to the west of the Panama Canal. To the east of Colón there are fringing reefs around Isla Galeta, but these have suffered considerably from pollution, and the adjacent mangrove areas have been devastated by oil spills over several years. More recently, however, there have been signs of limited recovery.

The best developed reefs lie in the direction of Portobelo, and particularly in the national park. Here fire corals, sheet corals, and knobby brain corals dominate the shallow water, then the reefs fall away in a gentle slope, with a mix of large boulder corals and soft corals, rarely extending below 15 meters.

Human impacts on the reefs are far greater in this region than elsewhere, and perhaps nowhere more than in the waters adjacent to the Panama Canal. In Bahía Limón, at the mouth of the canal, vast areas of coral reefs were removed to clear the channel, and to provide solid material for land reclamation schemes. It has been estimated that more than 50 million cubic meters of reef coral and other material were removed from the bay. Such coral mining for building materials is not unprecedented. There are records of large quantities of coral being extracted from the reefs near Portobelo, over a period of about 200 years, to be used in the construction of forts, churches, and other buildings across this fortified town. Low levels of coral

PEOPLE OF THE REEFS

The Kuna people are perhaps more intimately bound to coral reefs than any other people in the Caribbean. When Europeans arrived in this area the Kuna already had a complex civilization covering a wide area of eastern Panama and northwest Colombia. Since this time Kuna have maintained a clear and proud independence despite almost continuous contact with the European arrivals. Although there have been impacts, the Kuna have maintained many of their traditions, their language, and their way of life. At the start of the last century, in response to continuing abuse and infringements of their rights, the Kuna fought back and in 1930 their autonomy was recognized, with official status as an autonomous province granted in 1938. This status put the Kuna into a powerful, and almost unique, position compared with most of the Amerindians across both continents.

Traditionally the Kuna worked a form of shifting, slash-and-burn agriculture, but also fished in the rich coastal waters. From the 1900s more and more Kuna came to live in the coastal areas and settled on the nearshore islands. From these islands they still travel to the mainland to practice their agriculture, but they have increasingly come to rely on the ocean. The 41 inhabited islands are now densely packed – many have 300-500 inhabitants, but some have close to 5 000. As the populations grow, there is a tendency to physically expand the islands rather than move to new ones – they do this by infilling the shallow waters around their villages with coral and rubble mined from further afield.

Most of the fishing is for subsistence, using hand lines, but the Kuna also export high-value species – lobster, king crab, octopus, and conch, and occasionally groupers, snappers, and even turtles. While this fishing is undoubtedly having some impact, there are far more of these "target species" than may be found on most Caribbean reefs. Overall the impact of the Kuna on their coral reefs is still low, and their ownership of the adjacent watershed has meant that the forested hills have never been clear-felled. It is undoubtedly true that, without their presence, the reefs of San Blas would have suffered the same impacts from pollution, sedimentation, and rampant development that have devastated so many other parts of the Caribbean.

mining continue today in the building and protecting of the Kuna islands in the east.

Eastern Panama – the Kuna Yala

The Kuna Yala, or land of the Kuna, is an autonomous province in the east of Panama. Strung along this coast are 365 islands of the San Blas Archipelago, and all around are coral reefs. Arriving here by plane is a breathtaking experience. The sea is studded with myriad islands, and the waters flash and sparkle with patch reefs. Most of the islands are a rich green with a silver-white fringe of sand. Others are homes to the Kuna people, and on these it is almost impossible to see through the solid mass of thatched roofs to the island below. The shores seem to bristle with small jetties and dozens of canoes.

There are fringing reefs along much of the mainland coast and around the islands, with wide reef flats, then gentle reef slopes dropping down to about 15 meters. Hundreds of patch reefs are scattered across the lagoon. The outer edge of the lagoon also represents the continental shelf edge, typically only about 15-20 kilometers offshore. Here the outermost reefs have deep slopes carved into spurs and grooves by the waves. In the shallower water the movements are too rough for the majority of corals, and the most important species is coralline algae. Looking like a rather nondescript form of pink cement, this is actually a simple red algae which, to protect itself from grazing animals, lays down its own skeleton of limestone. In these areas of crashing waves, coralline algae are the major reef builders.

There are no residents in this region other than the Kuna, but there are a few hotels across the islands, supporting small-scale, low-impact tourism. The snorkeling is superb; however the only diving takes place from visiting boats, with permits, which come in from outside the region.

Within the protected waters of the lagoon there are extensive seagrass beds, while the shallow reefs are full of life. Damselfish dart and hover amidst these scenes, and pairs of butterflyfish loiter as they search for small crustaceans and worms to eat. The slender wrasses are abundant – apart from the well known bluehead and yellowhead wrasses, slippery dicks, clown wrasses, and blackear wrasses are all common in these shallow waters. Trunkfish are also often seen – among the strangest of this already strange group are the two species of cowfish which, in addition to their armor plated, box shaped bodies, are equipped with an extraordinary pair of sharp spines just above their eyes. These add to the already weighty defense provided by their body armor, making them a fairly indigestible mouthful for most predators. They are wary fish, and to get a look at their beautiful markings it is necessary to approach slowly and not to swim directly towards them – if they feel remotely threatened they will quickly swim off.

Snapper and grunts are common in all habitats across the archipelago. Jacks are widespread, including the yellow jack. Similar in shape to the bar jack, this species has a yellowish tint to its tail and around its eye, and can appear faintly yellow all over.

The Kuna people have maintained greater autonomy than almost any other native American people, and remain closely tied to their land and lifestyles – fishing, farming, and hunting.

Greater Antilles

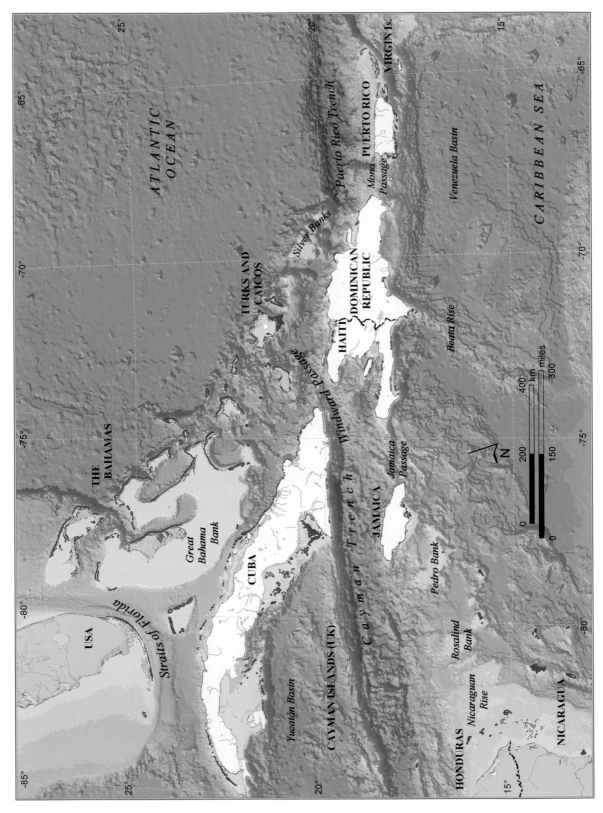

2.3 GREATER ANTILLES

In a great sweep tracing the northern border of the Caribbean Sea, the islands of the Greater Antilles encapsulate the full splendor of the region, both above and below water. These islands have their origins over 70 million years ago, when the underlying Caribbean tectonic plate pushed its way up against the North American plate – twisting and folding the rocks to form high mountains as it went. Even today many of those mountains remain. Cuba is a giant of an island, with high forested hills skirted by the longest coral reef systems in the Caribbean. Set against these like tiny specks of sand are the Cayman Islands. Rising from the depths of the Cayman Trench, they are, in fact, mountain tops, and the crystal clear waters all around abound in life.

Jamaica, Haiti, the Dominican Republic, and Puerto Rico continue this great island chain, rich with green hills and largely agricultural landscapes. Finally, at the eastern end, the Virgin Islands lie like a scattering of tiny jewels before the sea bed plunges down 2 000 meters into the Anegada Passage. Coral reefs lie all around these islands, although many have undergone the ravages of hurricanes and the abuse of humans. Most have suffered from overfishing, while pollution and sedimentation threaten the reefs closest to shore. Even amongst these islands, however, there are points of hope, signs of recovery, and places where beautiful underwater scenery will transfix the visitor.

Marine turtles were once plentiful all across this region. Their ancestors swam alongside dinosaurs some 200 million years ago, and turtles have swum in these waters all through the geological shifts and changes that drove the islands and seas of the Caribbean into their current positions. When Columbus first arrived in the Cayman Islands, green turtles were so numerous that his boats would have been clunking into their heavy shells as they were rowed ashore. Six species of turtle are found in the Caribbean – the loggerhead, green, hawksbill, Kemp's ridley, olive ridley, and leatherback. Decimated by centuries of overfishing, they have more recently been impacted by massive coastal development – the buildings and bright lights along beaches are a considerable disturbance to nesting turtles.

Thankfully a few turtles have survived, and still come to nest on the more remote and protected beaches, from the southern islands of Cuba to the undeveloped shores of the US Virgin Islands. Most countries have now given them some level of protection, and in a handful of places they are beginning to recover. There are few more entrancing sights than to come across some of these gentle, graceful creatures in their natural habitat, to watch them as they pick at sponges or seagrass on the sea floor, or when they swim slowly to the surface for a breath of air.

Quite apart from turtles, marine life through the Greater Antilles comes in a rich array. Wide areas have been set aside in parks and reserves. Some of these offer a taste of the Caribbean "as it once was", with fields of healthy coral providing a home to sharks and giant groupers. Everywhere, however, smaller life swirls in abundance, with the full complexity of sponges and tunicates, crustaceans, urchins, anemones, and snails – enough to challenge the best scientists and enthral the casual visitor.

A hawksbill turtle swims over the reef. Turtles are often highly inquisitive and regularly approach divers and snorkelers.

Western Cuba

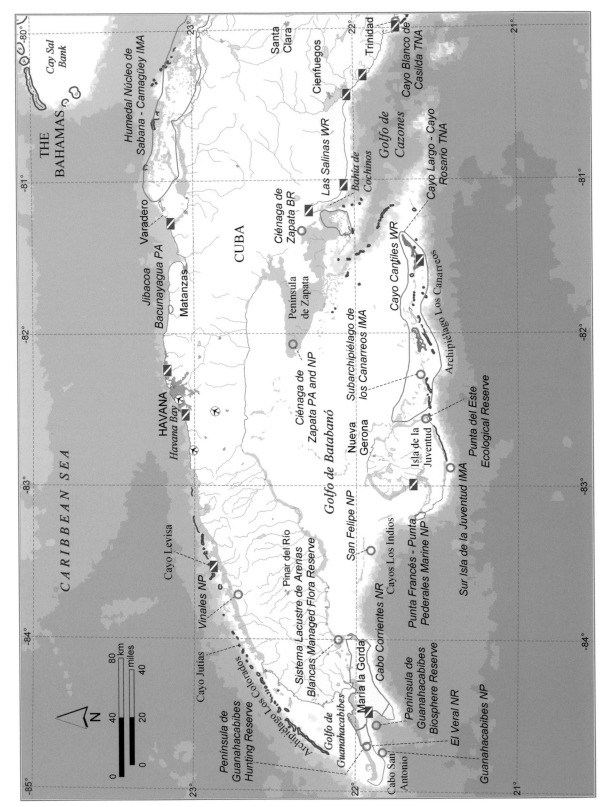

CUBA

Temperature	19-26°C (Jan), 24-32°C (Aug)
Rainfall	1 189 mm; peak rainfall Jun-Nov
Land area	110 437 km^2
Sea area	345 000 km^2
N° of islands	4 195
The reefs	3 020 km^2. Cuba has the largest reef systems in the Caribbean, running alongside chains of small islands quite far offshore. Away from these island chains there are nearshore fringing reefs. Most of the reefs are characterized by clear waters and dramatic walls. Corals are abundant and, in places, there are large aggregations of fish.
Tourism	1.8 million visitors. The majority of tourists go to coastal resorts, although visits to the cultural centers are very popular. Most diving is arranged from these resort areas and dive tourism is growing rapidly – there are 37 dive centers.
Conservation	Cuba has a strong interest in marine conservation and has made considerable efforts to protect at least some of its reefs. Overfishing is still a problem in most places, but there are areas of very healthy corals, and most reefs are still in a position to make a rapid recovery as protection improves.

The largest island of the Caribbean is also one of its great coral reef nations. Cuba is encircled by coral reefs, including the largest continuous reef system in the Caribbean. Many of these reefs are far offshore and relatively unaffected by pollution or sedimentation. While there is overfishing in some areas, large stretches remain in excellent condition, presenting a vibrant and critically important oasis of reef life at the heart of the Caribbean region.

Before the Spanish arrival, Cuba's native peoples probably numbered more than 100 000. They had a developed agricultural society, growing maize, cassava, beans, potatoes – and even tobacco and cotton. They also fished the surrounding waters. This society, like all others in the region, was devastated following colonization by the Spaniards, and within 50 years there were fewer than 3 000 native people.

As a Spanish colony Cuba grew relatively slowly, with a largely agricultural economy at its core. Cuba's struggle for independence in the late 19th century was interrupted by the Spanish-American War. This was followed by a long period of considerable US influence or control over Cuban politics. The island became something of a playground for the very wealthy, but behind this façade was a hidden majority living

Corals, seagrasses, and mangroves are often highly interlinked ecosystems, although it is rare to see such a close association as this, with healthy corals growing below (and even on) the roots of a mangrove in clear water.

Eastern Cuba

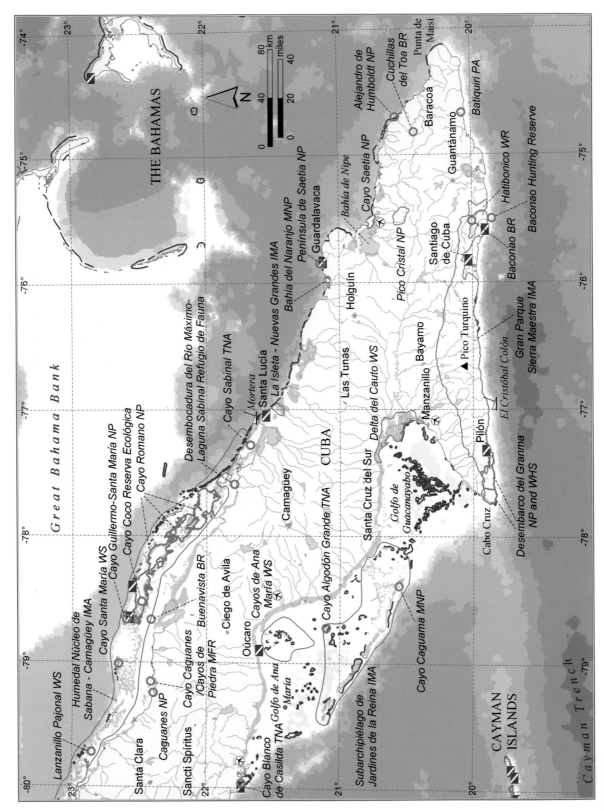

in extreme poverty. It was this inequality which led to the revolution which, in 1959, finally overthrew the old dictatorship of Batista. The new government gradually became more closely allied to the communist regimes of the then Soviet Union, and, throughout the Cold War, Cuba was relatively closed to the west. This was still a period of economic and industrial development, and indeed exploration of the marine environment.

Following the fall of communism in Europe Cuba underwent a major economic crisis, but this phase also saw the opening up of the country to international tourism. Although still closed to most visitors from the USA, Cuba now receives very large numbers of tourists from Canada and Europe. These visitors enjoy one of the most beautiful, culturally developed, and ethnically diverse countries in the Caribbean.

Cubans themselves have a close interest in the sea – with many enjoying swimming and snorkeling, although access to diving is limited. Fidel Castro himself was once a keen diver who said: "If I want a rest, if I want to relax, I go to the sea. I go to a small cay out there to scuba dive". Indeed his love of scuba diving is said to have led to some of the more extreme external attempts to assassinate the leader. These allegedly included a plan to give him a gift of a contaminated wet suit together with a contaminated snorkel tube, and another to pack a seashell with explosives and leave it in a place where Castro might find it while diving or snorkeling.

For visitors, the burgeoning tourism industry has opened up diving opportunities in resorts across the country. At the same time, large areas remain hard to get to, and indeed are largely unknown even to scientists. Low budgets mean that there are limited opportunities for Cuban scientists to work on the reefs even today.

The great Caribbean barrier reefs

A look at the map shows two quite distinctive types of coastline in Cuba. First there are the simple coasts, where deep water comes in very close to the mainland. Then there are shallow shores extending into wide lagoons and out to long chains of offshore islands. Coral reefs are found along about 90 percent of the coast, but are best developed off some of these island chains where there are nearly continuous shallow reef crests marked by a dramatic line of bright white surf.

These offshore barriers (sometimes called bank barrier reefs – see Chapter 1.1) make up the longest reef systems in the Caribbean and support the full spectrum of coastal life. There are extensive mangrove forests in the inshore areas and around some of the islands. The lagoons are filled with seagrasses, and there are occasional patch

The Malecon in Old Havana is a popular place for Cubans on a Saturday afternoon.

reefs. But the most extensive coral reefs are found on the outer edges of these shallow platforms where they have built up high reef crests, and then cover the rapidly descending reef slope beyond.

All of Cuba's reefs face out to deep, clear, oceanic water. In many places divers can swim along the top of a dramatic wall, which begins as shallow as 20 meters, and plummets to invisible depths. Work in submarines has shown that these walls often continue to 100-150 meters in depth, while dramatic inclines may continue to 600 meters or more.

Although they have not been as extensively studied as other areas, the lists of plants and animals observed in Cuban waters are already longer than for most other countries. Over 350 different fish are found living on or visiting the reefs (more than 1 000 marine fish have been observed altogether around Cuba). Almost all the Caribbean corals are represented, including some 50 stony corals and 55 gorgonians. Other studies have listed over 160 types of sponge and more than 300 kinds of macroalgae.

Schools of tarpon are a regular but nevertheless breathtaking sight on Cuba's reefs. Often reaching 1.5 meters in length, these fish have been recorded to 2.5 meters.

For many visitors this diversity may not be immediately apparent – all coral reefs seem to be a somewhat confusing tangle of life. Regular divers and snorkelers, however, will quickly start to make out things they have not seen before. In fact Cuba has many animals which are rare in other parts of the Caribbean. Whale sharks are regularly seen off the southern coast, and there are large and important stocks of groupers and snappers, including the rare jewfish. Manatees are not common, but are nonetheless found in a number of places, particularly in the shallow inshore waters.

Protecting the oceans

Being far offshore, many of Cuba's reefs are difficult to access, and are also buffered from the impacts of pollution and sedimentation, so many are in excellent condition. Sewage pollution is still a problem in a few places, while sugar and rum production, two of the country's major industries, are also responsible for the release of large amounts of nutrients. Although most of this pollution is biodegradable, it can lead to the removal of oxygen from the water and to localized deaths of fish and other organisms.

Fishing has not been a major part of the island's economy or history and, prior to the 1970s, was mostly coastal and on a small scale. But the 1970s and 1980s saw a rapid industrialization of the fisheries and many species became overfished, although not to the same level as in other places across the region. Unfortunately some spawning aggregations are still being targeted. Nassau groupers have declined considerably in number since the 1970s. Lobster fishing is probably still too high, and catches have fallen since 2000. Nonetheless, there are still more lobster here than in many other countries – indeed lobster remains one of the most valuable fisheries, worth an annual US$100 million in exports.

The control of fishing, like many other industries, is largely centralized, and this has enabled relatively rapid responses to environmental threats. Permits or licenses are required for most fishing. Conch fishing is now banned in many places and is strictly regulated elsewhere, so there are probably more conch in Cuban waters than in any other Caribbean country. Faced with threats to the lobster stocks, the number of licenses for lobster fishing has also been considerably reduced, while the minimum size for the mesh on fish traps means that most small or immature fish can leave the traps.

There is also a growing number of protected areas, used not only to control fishing but also as a tool for maintaining healthy reefs for a rapidly growing diving industry. For Cuba the value of marine resources for tourism is growing considerably, whilst the value for fisheries is stable or falling. There were an estimated 25 000 dive tourists in 2000, and this figure is expected to quadruple in the next few years.

ISLAND TOUR
The north

The westernmost tip of Cuba is the strangely shaped Península de Guanahacabibes. Much of this part is still wild and forested, lying within a protected biosphere reserve. Small numbers of divers visit the area, which has some of the healthiest coral populations in beautiful underwater scenery. The sea floor drops steeply into the open ocean close to the shore, and the rugged undersea walls are teeming with life. Star, brain, and sheet corals are interspersed with vividly colored sponges, including some very large barrel sponges. Blue chromis hover over the reefscape and angelfish are often seen making their way amongst the sponges. There are fewer large fish here than in other parts of the country, but tarpon are sometimes seen and jacks regularly sweep in past the wall. And in fact this area has the largest fish of all – whale sharks are regularly observed here from August to October. These gentle giants may reach over 12 meters in length, and live for over 50 years. They tend to move slowly through the water feeding on the plankton with their wide mouths agape – there can be few more thrilling experiences than a dive or a snorkel with a whale shark.

Along the northwestern coast, the Archipiélago Los Colorados is the smallest of the offshore island chains. It encloses the shallow Golfo de Guanahacabibes, a lagoon area which is mostly muddy, but also has patch reefs. Although not beautiful, these reefs are of considerable interest to marine biologists, precisely because they are growing in quite muddy waters (where corals do not normally grow). The surface of the reefs is largely covered by tube corals and the delicate ivory corals, as well as sponges, all of which are better adapted to these conditions. This is not a densely populated coastline and levels of fishing are also low. For the most part it is difficult for visitors to access, although a few tourists get to Cayo Jutías and Cayo Levisa.

Further east, the deep oceanic water again cuts in close to the mainland. While the waters around Havana are somewhat polluted, there are reefs as well as quite a number of shipwrecks lying not too far both to the east and west of the capital. Many Cubans enjoy snorkeling in this region (even along the waterfront in downtown Havana). The sea floor here is generally more gently shelving than elsewhere, with corals growing on small drops or on banks rising above their surroundings. Fishing pressure is relatively high, so larger creatures are not so common; however there are healthy corals and sponges, as well as small fish in abundance.

Archipiélago Sabana-Camagüey

Running along much of the central coast of northern Cuba is the longest reef system in the Caribbean. The Archipiélago Sabana-Camagüey

Rich dense fields of staghorn coral on the "back reef" – the first to go in polluted waters – are still common in Jardines de la Reina.

The jewfish is the world's largest grouper, an awesome fish which is naturally inquisitive and will often approach and hang around divers.

spans five of the country's 14 provinces, stretching some 465 kilometers from east to west. The western part is sometimes referred to as the Archipiélago de Sabana and the eastern part as the Archipiélago de Camagüey.

The mainland along this entire coastline is low lying and mangrove fringed. Offshore is a shallow muddy lagoon, typically only 2-3 meters deep. There are wide areas of mangrove "islands" – whether they are islands or not is a matter of opinion, as many do not have any dry land at all at high tide. These forests of the sea are usually teeming with small fish, which begin life here before migrating out to live as adults on the reef itself. The outer edge of the lagoon ranges from 6-35 kilometers offshore and is marked by a chain of larger islands. Beyond these lies a reef crest, with areas of bare rock colonized by occasional gorgonians and patches of elkhorn and staghorn coral. The waters then gradually deepen across the reef slope to the top of a wall, typically at about 30 meters from the surface.

BONEFISH AND FLY FISHING

The bonefish is a pale silvery blue fish with a strange conical snout and downwards pointing mouth. It is not a reef fish, but is often found in nearby shallow water flats, amongst mangrove forests, over seagrass beds, and in river mouths where it feeds on worms, mollusks, and crustaceans. Like tarpon, bonefish are considered to be an ancient form of fish, with fossil forms dating back some 120 million years.

They are highly sought after by fly fishermen, and many regard the bonefish, alongside tarpon and permit, as the ultimate game fish. The challenge of catching these fish includes the considerable skills of placing a "fly" and hooking the fish. Although not large (reaching 80 centimeters), they fight hard once on the hook.

The sport is in the catching, and these fish do not make good eating – they get their name from the large numbers of fine bones in their flesh. Once caught they are typically returned to the water and so, even though these fish are not common over wide areas of the Caribbean, fly fishing does not pose a threat. In fact the reverse is true – fishers are willing to spend large amounts of money to fish in remote and pristine areas, notably in Cuba, Belize, and the Bahamas. Their interest is provoking greater efforts to protect mangroves, seagrasses, and adjacent coral reefs in many places, as it is in everyone's interest that such places remain wild and healthy.

In the far west this archipelago is almost connected to land by the long, narrow Península de Hicacos, better known as Varadero. This is home to the biggest concentration of beach tourism in Cuba. Diving here is mostly from boats which go out to the northeast to see healthy concentrations of coral, particularly soft corals. Small fish – chromis and blue tangs – dart amongst the sea rods and plumes, while the occasional tiger grouper may be seen. Grunts and schoolmasters loiter in quite large numbers, while higher in the water there are often Bermuda chub and Atlantic spadefish. A number of wrecks, deliberately sunk here in the late 1990s for divers, are beginning to attract marine life and will be smothered within a few years.

Diving is also popular at the newer resort areas around Cayo Coco, the first of a series of much larger islands towards the eastern end of the archipelago. Although parts of the island are quite developed, wide areas are still forested, and pelicans and other seabirds are frequently seen. Underwater, the massive dome-shaped star corals, which are important reef builders, are common. Fishing pressure is not high here and so divers regularly see tarpon, larger groupers and snappers, and even eagle rays, southern stingrays, and nurse sharks. A close look amongst the soft corals is also likely to be rewarded by the occasional balloonfish. These extraordinary creatures wait in the shadows until night-time when they head out to feed on a variety of invertebrates such as mollusks, hermit crabs, and even sea urchins. Up close you can see that they are covered with fine spines, and, rather than retreating when threatened, these strange fish inflate themselves with sea water so that the spines point outwards like a pincushion. Some divers harass these fish in order to provoke such a response, but they should be discouraged – it is a reaction to extreme stress, and handling can damage the fish's skin.

The final diving center along the archipelago lies where the reef system rejoins the mainland, around the resort of Santa Lucía. Here, once again, are coral walls which abound with life, including sponges and soft corals. Smaller fish such as hamlets and butterflyfish are common, but also groupers. In the deeper water, eagle rays are regularly seen, and, occasionally, manta rays. Near here is the wreck of the steel merchant ship, the *Mortera*, which sank in 1896. She had been damaged in a collision with another ship, but carried on, going down here as she tried to make for cover through the Nuevitas Channel. The surface of the wreck is smothered in life, and bull sharks are sometimes seen nearby.

The east

The eastern end of Cuba, from Santa Lucía to the eastern tip of Punta de Maisí and along the south coast back to Cabo Cruz, is the most mountainous region of the country, and the beautiful, dramatic scenery onshore is mirrored by a steep wall of coral offshore. The highest point of the Sierra Maestra, Pico Turquino (1 972 meters), lies close to the south coast where, just offshore, the Cayman Trench plunges to a depth of more than 7 200 meters – the distance from ocean floor to mountain top is greater than the height of Mt Everest. Along much of this coastline there is still a shallow platform close to the shore, and the fringing reef can often be seen as a line of breaking waves some 100-200 meters offshore. This shallow reef crest is built largely from staghorn coral and branching fire coral. Where these are still living they can make for beautiful snorkeling on calm days. Beyond the reef crest the water slopes to a depth of about 30 meters before plunging down in nearly vertical walls.

With the mountains so close to the reefs, underwater visibility can be poor during the wet season (June to November); however during the dry season this area is one of the drier parts of the country. Reefs are widespread along the entire coast, and diving is popular from centers in Cuba's second city, Santiago de Cuba, and from the northern resort of Guardalavaca. Perhaps the best known wreck in the country is the *Cristóbal Colón*. She was once the pride of the Spanish Navy, but was attacked during the Spanish-American War in 1898, when a small fleet tried to escape through an American blockade of Santiago. The other vessels were all quickly overpowered, but the *Cristóbal Colón* broke through, making some 80 kilometers before the US fleet caught her, when she

surrendered and ran ashore. Badly damaged, she could not be kept afloat and sank on a sandy slope. Today she has become a wonderful artificial reef, teeming with life. Small schools of giant, bright silver tarpon are often seen around the wreck.

The hutia, a large rodent, is widespread on some of Cuba's offshore island chains as well as in the Zapata swamps where, among other things, it feeds on mangrove leaves.

Archipiélago Jardines de la Reina

The second longest reef in the Caribbean runs from Cabo Cruz to the town of Casilda. This reef and its islands make up the Archipiélago Jardines de la Reina, enclosing the Golfo de Guacanayabo and the Golfo de Ana María. In many ways the barrier reef here is very similar to the Belize Barrier Reef and to many of the barrier reefs of the Indian Ocean. In places it is over 80 kilometers from the mainland, and there is a lagoon reaching to a depth of 30 meters between the reef and the shore.

The lagoon floor is muddy nearest the mainland, becoming sandy further offshore with extensive seagrass. However, this is one of Cuba's major prawn trawling grounds, and wide areas of seagrass have been plowed up by this destructive form of fishing. In a few places in the lagoon there are patch reefs.

On the lagoon side of the islands there are many back reefs scattered amongst extensive seagrass beds. These are dominated by branching corals and can cover wide areas with 100 percent coral cover, making for very beautiful snorkeling. The islands themselves are largely mangrove fringed, and some have no dry land at all.

A large section of the reef is in a marine park

– visitors are charged to fish or dive here, and there are a number of restrictions. There is only one diving operator, on a floating hotel moored between mangrove islands. No commercial fishing is allowed other than a small licensed lobster industry. The coral reefs here offer a rare opportunity to see how the Caribbean once was, with abundant fish, healthy coral populations, and all of the related habitats of back reefs, mangroves, and seagrasses.

Groupers abound, including sizable numbers of black and yellowfin groupers, and there are even considerable numbers of jewfish, with individuals weighing up to 200 kilos. These are the largest bony fish to be found on coral reefs anywhere in the world. Their inquisitive nature means that they often come to find divers rather than *vice versa*. Sadly, this lack of fear has contributed to their virtual disappearance from many other reefs across the Caribbean as they have fallen prey to spear fishers. Large snapper such as cubera and dog snapper are also abundant, and visitors will occasionally encounter turtles. Silky and Caribbean reef sharks are resident at a couple of sites as a result of occasional feeding by dive operators. The shallow flats here are highly regarded by fishermen for their bonefish, tarpon, and permit.

Golfo de Cazones

To the west of the Golfo de Ana María, the waters of the Yucatán Basin cut in close to the mainland in the deep Golfo de Cazones. Although not marked on many maps, coral reefs fringe much of this coastline, sometimes only 100 meters or so offshore. In a few places there are opportunities to snorkel over beautiful coral gardens where elkhorn and fire corals are abundant, along with the common fan corals. Look out for the beautifully patterned flamingo tongue snails on gorgonian corals. They actually feed on these corals, but during the day are usually found near the base of the fans.

In general these reefs have good coral cover, and are populated by large numbers of smaller fish such as blue and brown chromis. Larger fish are not so abundant, but turtles, eagle rays, and nurse sharks can all be seen here. Rope sponges and whip corals abound on the steeper walls.

Fringing reefs continue, even into the Bahía de Cochinos. This is the infamous Bay of Pigs

Green morays are common on the reef slopes, here resting in a hole surrounded by rope sponges.

where CIA-trained Cuban emigrés attempted to attack the new revolutionary regime in Cuba in 1961. Although there are two wrecks from the invasion in the bay, they are too deep for divers to access.

Archipiélago Los Canarreos and Isla de la Juventud

Offshore and to the west of the Bahía de Cochinos the coastal shelf rapidly widens into the vast shallow area of the Golfo de Batabanó. Despite its vast size, this gulf is rarely more than 5 meters deep. Much of the bottom is sandy or muddy, with seagrass over large areas. There are patch reefs in the eastern part of the lagoon, but the greatest concentrations of reefs are along the outer edge. Huge volumes of water flow on and off this shelf with the changing tides and there can be quite low visibility as a result, particularly near the channels that cut through the reef and on to the shelf.

Before describing the coral reefs found here, brief mention should be made of the Zapata Peninsula, which juts into the eastern side of the Golfo de Batabanó. This peninsula includes one of the largest single blocks of mangrove forest in the entire Caribbean, as well as wide areas of freshwater swamps and grasslands. A protected area with international recognition and status,

Fire coral, as the name implies, can inflict a minor, burning sting and leave a rash, often for several days – but they are also responsible for building some beautiful coral gardens in quite shallow water.

Zapata is home to numerous species including the Cuban crocodile, the American crocodile, and large numbers of Caribbean flamingos. Some of Cuba's unique bird species are also found here, including the Zapata rail, Zapata wren, and even the world's smallest bird, the bee hummingbird, which measures only 5 centimeters and weighs some 2 grams.

Offshore in the east, a chain of hundreds of low islands, tiny islets, and mangrove cays makes up the Archipiélago Los Canarreos. The largest of these, to the east, is Cayo Largo – a tourist center and popular diving destination. Here the coral walls begin in quite shallow water. Tarpon come on to the reefs, and green and hawksbill turtles are also quite often seen. In many places on the shallower reefs lane snappers and schoolmasters join the usual dense crowds of grunts waiting for nightfall. Yellowtail snapper are also abundant high above the reef – unlike other snappers these inquisitive fish generally feed in mid-water, opportunistically taking small animals from the plankton and occasionally catching small fish.

The Golfo de Batabanó itself is one of the largest lobster fishing centers of the world and lobsters are also regularly found on the reefs. Like many crustaceans, lobsters are nocturnal, so during the day they usually lurk in dark recesses with their long antennae pointing outwards to sense danger. At night they scour the reef for mollusks, worms, and small crustaceans.

The Isla de la Juventud is Cuba's largest offshore island, a low-lying agricultural landscape. Reefs fringe most of the south coast, but are best known around the diving center to the west. Here a great diversity of corals has built some spectacular scenery, with towers and gullies, and even long caves – with every surface thickly encrusted with lettuce, brain, and star corals, and with tube and rope sponges. Black coral is abundant on deeper slopes. Groupers are relatively common here, including the Nassau grouper, and there are also larger snappers, barracuda, jacks, and angelfish, as well as hordes of smaller species. Moray eels often lurk amongst the corals. Up close blennies and gobies are always widespread, and there are many Christmas tree worms poking their filtering apparatus out from the surface of brain corals.

Corals continue along the shelf edge offshore from the Cayos Los Indios and San Felipe until they rejoin the mainland, but these areas remain little known to scientists and largely inaccessible to divers. Like so many other areas in Cuba, it seems likely there could be even more undiscovered marine treasures in these waters.

CAYMAN ISLANDS

Temperature	18-30°C (Feb), 23-33°C (Jun)
Rainfall	1 435 mm; peak rainfall May-Oct
Land area	277 km^2
Sea area	119 000 km^2
N° of islands	3
The reefs	230 km^2. Reefs fringe most of the islands, often not far offshore, and there are dramatic walls plunging to great depths. Groupers, rays, large pelagic fish, and sharks are abundant.
Tourism	334 000 visitors plus 1 215 000 cruise ship passengers. There are 31 dive centers. The vast majority of the population and tourist facilities are on Grand Cayman; the other islands are quiet and unspoiled.
Conservation	The importance of the reefs is well recognized and there are growing efforts to protect reefs in reserves, as well as to manage fisheries and minimize the negative impacts of tourism.

Tiny, and almost invisible on many maps, the Cayman Islands are fabulous gems. Although low lying, these three islands are actually the high points on a vast underwater mountain chain, the Cayman Ridge. To the south the sea floor plunges down 7 535 meters (4.7 miles) in the Cayman Trench, the deepest point in the Caribbean Sea. To the north, too, the waters quickly descend into the deep dark depths of the Yucatán Basin.

Of course fishermen and visitors barely skim the surface of such deep waters, but they still reap the benefits. Creatures usually found out in the open ocean, including tuna, sharks, and manta rays, come close to land. The deep water constantly refreshes and replenishes the reef environment – with little chance of the waters becoming polluted or murky.

For divers and snorkelers the waters are incomparable. In many places vertical walls rapidly descend into deep, dark blue. Caves and canyons abound.

Cayman Islands

The islands are themselves built from ancient coral reefs which grew here tens of millions of years ago. Between the long, white sand beaches, much of the coastline is a jagged limestone rock, and if you look closely you can see the fossils of corals and shells in the cliffs. Like all limestone, the rock is porous, so there are no ponds or rivers on the islands. Instead, the rainwater percolates through and "floats" within this rocky matrix on the seawater below. Wells dug through the islands come first to freshwater, reaching saltwater if they go deeper.

Unlike the other islands of the Caribbean, the Caymans were never inhabited by native American peoples and Christopher Columbus was probably their first visitor in 1503. Some time later, an account of Sir Francis Drake's 1586 voyage describes how the fleet came upon "…two islands called Caimanes, where we refreshed ourselves with allagartas [crocodiles] and greate turtoises [turtles], being very ugly and fearefull beasts to behold, but were made good meate to eate". The name Cayman Islands comes indeed from the great abundance of crocodiles (which the Carib Indians elsewhere called *caymanas*) which lived here until hunted to extinction.

The islands today

The Caymans are now a British Overseas Territory and maintain a strong British culture. Tourism is the major industry on the islands, with 1.5 million visitors a year. Diving is one of the most popular activities, and almost half of all visitors venture on to the coral reefs. Financial services are the other mainstay of the island economy – tens of thousands of companies and several hundred banks use the islands as a confidential tax haven.

Almost the entire population lives around the capital George Town, at the West End of Grand Cayman. Most of the tourists stay here too, in visiting cruise ships or at a string of hotels running along Seven Mile Beach. Despite the development, healthy reefs teeming with life surround the island. Likewise, the shallow lagoon of North Sound at the heart of the island still contains extensive mangrove forests – tremendously important for many coral reef fishes which breed and live as juveniles in these protected waters.

TURTLE STORY

It is hard to imagine what the underwater world of the Caymans was like when Christopher Columbus arrived there, but we have a few clues. The islands were briefly named Tortugas, after all the turtles. Columbus' son, Ferdinand, wrote in their log of 1503 that the islands were "full of tortoise [turtles], as was the sea about". Much later the British harvested the green turtle nesting grounds of the Caymans to supply their colonies in Jamaica. Between 1688 and 1730 they took about 13 000 turtles per year. It has been estimated that, before this catastrophic harvest, the total turtle population around these tiny islands was over 6 million animals. Scientists are struggling to imagine what impact 6 million feeding turtles might have had on the reefs and lagoons, but if people could have snorkeled at those times it would have been an awesome sight, with so many turtles that it would have been difficult even to see the reef!

Although only a few turtles remain, the Caymans is still one of the better places to see these magnificent creatures. They are often inquisitive and will come quite close to divers, but should not be harassed.

The final part of the turtle's tale in the Caymans is told in the Cayman Turtle Farm on Grand Cayman. This is the most comprehensive facility of its kind anywhere in the world, and has been breeding turtles since 1968. It holds a stock of about 350 breeding turtles, but at any one time there may be 10 000-20 000 turtles at the farm. The priority is to rear turtles for meat; however the farm also releases large numbers of one-year-old turtles into the wild. Over 30 000 have been released so far. If you see a turtle while diving or snorkeling have a look for a distinctive tag and markings on its shell which will tell you if it came from the farm. The turtle farm breeds mostly green turtles, although it has also had some success with breeding the very rare Kemp's ridley turtle (pictured above).

Today then, the Cayman Islands are the only place in the world where it is possible to eat turtle meat without either breaking the law or threatening a species with extinction. It remains illegal to export meat, or indeed any part of a turtle under international law. Tourists should be aware of this and should refrain from taking souvenirs made from turtle shell home with them.

The two smaller islands are often called the Sister Islands. Cayman Brac has only a small population, while Little Cayman is almost uninhabited, with fewer than 100 permanent residents. "The Brac" is mostly surrounded by rocky coast, with waters plunging to great depths just offshore. Little Cayman is less rugged, with shallow coastal waters in places. Inland there are swamps and some mangrove forests.

The state of the reefs

The value of the coral reefs to the Cayman islanders is immense. This has been recognized and over one third of all the shallow waters is

Sister Islands

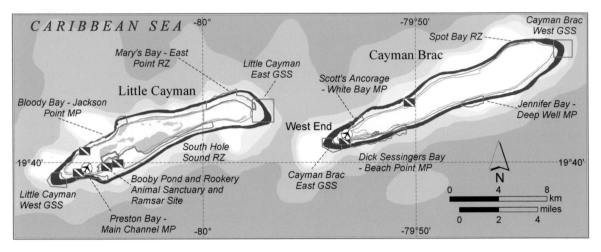

Grand Cayman

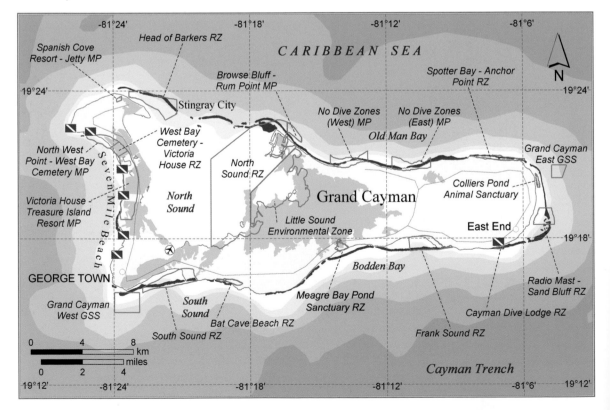

protected. Most fishing is banned in marine parks and replenishment zones. Over 250 permanent mooring buoys have reduced the need for anchoring on the reefs and these also encourage dive operators to disperse more widely among the reef areas.

Nonetheless, in a few places the coral reefs have suffered. Off George Town the rapid growth of the town and tourist facilities, together with physical pounding by boat anchors (notably from cruise ships) has damaged many reefs. Overfishing has reduced fish numbers in parts of Grand Cayman and there is also a small traditional harvest of turtles, which, although strictly licensed, is probably damaging.

Many still regard the Caymans as offering the best diving in the Caribbean, and continuing efforts to protect these reefs should ensure that this situation holds sway in the years ahead.

ISLAND TOUR

Fringing the shore of each island there is typically a shallow reef flat or lagoon, reaching a meter or two in depth. This is made up of sand and rubble, but is often overgrown with large meadows of seagrass. Beyond this lies the reef crest, where waves break over the shallowest rocks and corals. Below the reef crest elkhorn and staghorn corals can be abundant, and beyond these lies a largely sandy slope interspersed with coral outcrops.

Deeper down, rocks and corals form a more regular pattern of spur and groove formations – high ridges of coral rock interspersed with narrow sand-filled channels. Deeper still this dramatic scenery gives way to steep reef slopes, and, in many places, vertical walls dropping to depths of hundreds of meters. These walls are one of the most characteristic features of the underwater world of the Caymans – in Bloody Bay Marine Park on Little Cayman the wall drops rapidly to nearly 2 000 meters. Star, brain, sheet, and lettuce corals are abundant, and deepwater sea fans are found on the steeper sections, particularly where there are strong currents. In many places the scenery is further enhanced by deep canyons or caves; small holes which divers can swim

Deep fissures, canyons, and caves intersect the reef slope and walls in the Cayman Islands – many were formed during the ice ages when these reefs were dry land.

The Nassau grouper is a solitary predator. Still common in the Caymans, these magnificent fish are considered commercially extinct further east in the Greater Antilles.

through; and dark recesses with large fish lurking at their entrances.

The bigger fish, including groupers, snappers, and parrotfish, are still common throughout the Caymans, but particularly in some of the marine parks and reserves where there is little or no fishing pressure. The Sister Islands offer some of the best opportunities for seeing fish – the sheer numbers as well as the very large size of individuals never fail to impress. In a few places eagle rays make a regular and thrilling appearance, sometimes in groups forming striking "aerial" displays.

STINGRAY CITY

Among the most famous underwater scenes in the Caymans are those found in Stingray City. Here, in the northwest of Grand Cayman, large numbers of stingrays have become used to people, and are regularly fed by divers.

Visitors are taken to a wide sandy plain, in shallow water where they can snorkel or dive. Within minutes of arriving, the impressive dark forms of stingrays arrive, flying gracefully through the brilliant turquoise water, to see if there is any food. The rays have been approaching boats in this location for years: even before divers and tourism, fishermen used to come here to clean their catches and the rays long ago learned to associate people with a free meal.

If you look on the base of a stingray's tail you can see a short spine – not a true "sting". It is only ever used for defence: if they are caught or trodden on, stingrays will flick up their tails and the spine can inflict a nasty wound, but if left unmolested they are totally harmless.

Around some of the larger coral heads, silversides gather in spectacular schools like shimmering clouds.

One grouper, the beautiful Nassau grouper, is worthy of particular notice. Once a year, around the full moon in December or January, these fish gather from many miles around to spawn in vast numbers. In 2002, over 5 000 were seen at one time at a spawning site in Little Cayman. Sadly, when fishermen discover these spawning sites they can quickly decimate an entire population. Nearly 2 000 of the fish observed at that one site in 2002 were captured. This sort of fishing has driven these fish close to extinction in many parts of the Caribbean. However, new legislation was passed in the Caymans in 2003 to protect all the main spawning sites, so there may now be a chance of recovery.

Sites facing deep water, particularly in protected areas or near strong currents, are the best places to see big schools of predators, and also occasional visitors from the deep such as sharks and wahoo. The northern shores of all the islands have some of the best opportunities for these "big fish" dives.

Sharks are regularly seen in the waters of the Cayman Islands. In early 2002 the government introduced a ban on shark feeding. While many visitors to coral reefs love to see sharks, artificial feeding can bring problems as the behavior of sharks alters and they become less wary of people. Many believe that this increases the risk of shark attacks. Even if this is not the case it is clear that feeding influences the natural behavior of these powerful and beautiful creatures (see page 73). Elsewhere, the feeding of other fish still goes on. This can make for exciting diving or snorkeling, but it should be remembered that it is impossible to watch the natural goings-on of the reef in the middle of such a frenzy.

As if the reef walls, big fish, and tightly packed corals were not sufficient, the Caymans also have a few wrecks, including some deliberately sunk for divers. These include the deep *Oro Verde* on Grand Cayman and the Russian MV *Capt. Keith Tibbetts* in Cayman Brac. They offer a wealth of shelter for moray eels, snappers, and horse-eye jacks, while garden eels can be seen in the surrounding fields of sand.

JAMAICA

Temperature	Montego Bay: 22-27°C (Jan-Feb), 25-31°C (Jul-Aug)
Rainfall	1 280 mm; wetter May-Nov; south coast generally drier and a little hotter
Land area	11 044 km²
Sea area	251 000 km²
N° of islands	47 (1 large)
The reefs	1 240 km². Many of Jamaica's reefs have been degraded, but there are still some areas of beautiful underwater scenery with a profusion of sponges, coral, and algae. Far offshore to the south are shallow coral banks, still rich with reef life, but difficult to access.
Tourism	1.3 million tourists plus a further 840 000 cruise ship passengers. Many come for the beaches and the culture, and access to the coral reefs is easy and popular. There are 36 dive centers, which take visitors to the best reefs.
Conservation	Important new efforts are driving considerable changes in the marine parks around the island and it is likely that these will lead to increases in fish numbers and recovery of corals. Elsewhere the reefs are in a poor state.

The third largest island of the Caribbean, Jamaica is a beautiful, hilly land swathed in green fields and verdant forests, with steep valleys and waterfalls. The island takes its name from the original Arawak inhabitants, whose word *Xaymaca* means "the land of wood and water". The island is also alive with a rich culture and has a broadly based economy. Although tourism is one of the most important industries there are many others: sugar, bananas, rum, and coffee dominate the agricultural scene, while there is some mining and there are about 20 000 fishers operating across Jamaica's wide ocean territories.

Jamaica's coastline is highly varied with sheltered beaches, rocky shores, and occasional mangrove forests. There are several rivers, and the river mouths and adjacent mangrove areas are still home to a few American crocodiles and even manatees. The north coast is quite rugged, with hills coming right up to the shore, and deep waters lying not far beyond. To the south there are wider shallow plains extending out across the South Jamaica Shelf. Far out from the island of Jamaica are some tiny coral cays on Morant and Pedro Banks, with rich coral and fish communities all around.

The island of Jamaica was inhabited by Arawak people from at least AD 650. Christopher Columbus landed in 1494 and European arrival was rapidly followed by the demise of the Arawak from European diseases and fighting. Spanish rule was overthrown by the British in 1655, and British rule began with years of piracy, often informally supported by Britain. One pirate, Henry Morgan, was eventually knighted and made governor of the island.

Many international travelers favor Jamaica for its beaches, its landscape, and its culture, and a highly developed tourist infrastructure has grown along the west and north coasts. In all of these tourist centers there are opportunities to visit the reefs for diving, snorkeling, and,

Yellowline arrow crabs are abundant throughout the Caribbean and beyond, but are often overlooked.

Jamaica

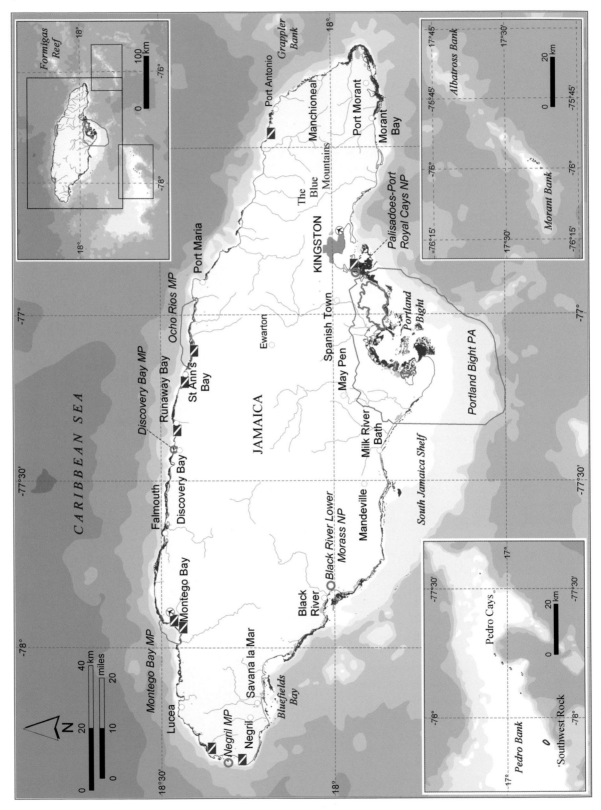

for those who prefer to remain dry, even visits in glass-bottomed boats. Underwater, many of Jamaica's once famous coral reefs have declined over the last 20-30 years (see box). However, there are still some fascinating and beautiful reefs, and efforts to protect some of these are beginning to show positive effects. Marine parks have been established, usually in close consultation with the local populations, and restrictions have been placed on certain fishing activities. It is likely that visitor fees will be introduced in the near future to help fund better patrolling of their waters.

Hurricanes pulverized the corals off Jamaica in 1980 and again in 1988, after which recovery remains limited.

ISLAND TOUR

The westernmost part of the island around Negril has rapidly grown from a small village to a major tourist center; however there have been strenuous efforts to protect the coral reefs. Below water the scenery is gentle, with reefs rising out of a gently sloping sandy shelf. Sponges are abundant, including the large orange elephant ear sponge – these giant sponges can encrust the reef, but often grow out in thick leathery folds, their surface covered in fine pits and grooves.

In the shallower waters, sea rods and sea plumes are common, with bicolor damselfish and occasional groups of goatfish moving about the scenery. Balloonfish are sometimes seen amongst the corals and crevices, and trumpetfish are often found. The latter are highly successful predators – they make full use of their extraordinary shape, sometimes shadowing other fish, sometimes holding themselves still alongside tall sea rods. When prey such as small damselfish are within range they lunge forwards and engulf them with their enormous mouths.

Looking up close at the reef, and especially among sponges, arrow crabs are abundant. Just a few centimeters across, these creatures appear to be all legs and no body. The body is a tiny triangle drawn up into the most extraordinary pointed snout with eyes sticking out like little balls. These strange creatures are generally unafraid of divers. They feed at night, and up close it is possible to see that the shorter front two legs are tipped with very delicate purple pincers which they use to scavenge for food and

CYCLES OF DECLINE

Jamaican coral reefs have suffered more than many others in the Caribbean, but they are also among the best studied, with many scientists seeing them as a microcosm of the problems facing coral reefs around the world. Overfishing has long been a problem here, even since the mid-19th century, when the colony was already importing 85 percent of the fish it needed to feed its rapidly growing population. By the middle of the 20th century, studies showed that few fish were surviving more than a year before being caught – and very few were surviving long enough to breed.

Despite this single problem, corals continued to flourish until 1980, when Hurricane Allen crashed over the reefs with waves reaching over 11 meters (35 feet) in height. Hurricanes are natural events, but their impacts can be devastating – reducing wide areas of coral to rubble in hours. Slowly the corals began to recover, but then a second disaster struck. In 1983 the resident long-spined sea urchins began to die in vast numbers from an unknown disease (see Chapter 1.2). These urchins graze on algae, and in Jamaica – where other grazers such as fish and conch were so heavily overfished – they had assumed a critical role in keeping algal growth down.

The algae, perhaps benefiting from higher levels of nutrients in the water from sewage pollution, proliferated and their dense mats and wafting fronds meant there was little space for new corals to settle, while existing corals were only just managing to survive. Then, in 1988, Hurricane Gilbert hit the island and once again the corals were smashed by the waves. This time, however, the fast-growing algae had a stranglehold and there was no space for new corals.

In the 1960s, algae covered only about 4 percent of the coral reefs, but now they cover about 90 percent of most Jamaican reefs. Scientists are worried that it will be difficult for the corals to re-establish. Perhaps by rigorously preventing fishing in at least a few areas, grazing fish may return in sufficient numbers to reduce the algae. Then at least there will be a chance for new corals to settle.

to pick at the filters of feather duster worms. These tiny creatures are truly widespread in many habitats across the Caribbean and beyond, and have been found at depths of 1 500 meters.

The northern shore of Jamaica has, or had, fringing reefs along much of its length, and the waves breaking over the reef crest are still clearly visible from land. But where there were once wide fields of coral, today the shallow waters are dominated by algae and rubble. There are still a few good places for snorkeling, but most wide areas of healthy coral are now found in deeper waters, the realm of divers.

Montego Bay Marine Park is one of the main dive centers, and also offers opportunities for newcomers to get close to the reef on glass-bottomed boat or snorkeling trips. Here, and along much of the northern coastline, there are steep or vertical walls of coral rock, heavily overgrown with sponges, wire corals, and deep sea fans. Here too are caves and crevices which divers can explore – glassy sweepers often hover in the dark, and with a torch divers can observe the fantastic colors of sponges and zoanthids which carpet the otherwise dark walls. Out into the blue water beyond the reef, the slender silver bullets of cero, or occasional schools of horse-eye jacks, make a regular appearance.

In a few deeper places there are still some fine colonies of finger corals, forming dense stony bushes, and the long wiry arms of brittle stars can sometimes be seen twisting amongst the branches. Big groupers have largely gone from Jamaica, but graysby and coney are still wide-spread. Both of these small groupers are terri-torial, with single males defending a patch of reef and a harem of two to five females.

One of the most important marine lab-oratories in the Caribbean was established at Discovery Bay in 1965. Just to the east, the popular tourist areas of Runaway Bay and Ocho Rios take visitors to an almost continuous string of coral reefs dominated by steep walls. Up close on the reef slope divers will see the dazzling fairy basslets and the occasional barred hamlet or shy hamlet. It is also quite easy to find giant sea anemones, which capture small crustaceans and other food with their powerful stinging cells. Some creatures have learned to live with them,

ENJOYING ALGAE

Many divers tend to overlook algae, present on every reef but often considered a poor second to the graceful plumes or giant structures of corals. In Jamaica, where the algae have somewhat overtaken the corals, there are ample opportunities to look a little closer. A lot of simple algae just look like greenish or brownish fuzz growing on the rock in dense mats – these may not be much to look at, but they are the favored food of parrotfish, surgeonfish, and damselfish. Much larger algae include the graceful curving folds of white scroll algae and the bright green tufts or bushes of *Halimeda*. The latter is a group of species, sometimes called hanging vine or watercress algae, with fronds made from numerous, small, disc-shaped "leaves". Each leaf is quite firm as these algae secrete calcium carbonate, rather like corals, and they have an important role to play in the creation of coral reefs and sandy beaches. A much stranger algal form are the shiny, greenish blue balls known as sea pearls which are found moored to the sea floor. Reaching 5-6 centimeters in diameter, these are remarkable as being among the largest single-celled organisms in the plant or animal world.

Algae tend to thrive in areas of slightly higher nutrients than coral. On reefs they are often kept in check by grazing fish, but some damselfish actually encourage algae. The yellowtail damselfish, with its bright tail and vibrant blue pinpricks of color on its dark flanks, is one such "farmer fish". Watched for a while, each fish can be seen stoutly defending a tiny territory – its farm – within which the brown algae has grown into thick mats. In tending their crops they will see off quite large parrotfish, and will even try to chase off divers who come too close.

Green algae dominate many of the reef scenes in Jamaica.

and it is always worth making a close inspection to spot the translucent form of Pederson cleaner shrimps, or the tiny, polka-dotted squat anemone shrimp.

In the open, schools of creole wrasse can often be seen moving around the reefs, while a little higher above the corals, fast-moving schools of slender, metallic blue boga are regular visitors. Both species feed on plankton in the open water. Also out in the open water, especially close to the surface, floating gelatinous shapes can often be seen. Jellyfish, such as moon jellies, are familiar to most people, but there are many other strange, near invisible creatures to be found. These include comb jellies, roughly oval in shape and up to 8 centimeters in diameter, with complex ridges and folds in their delicate bodies. The most visible parts of these ghostly animals are lines of sparkling iridescence formed by cilia – microscopic hairs that beat in order to swim slowly through the water.

Towards the eastern end of the north coast, Port Antonio provides another point of access to the coral reefs. While staghorn coral is making a comeback in a few shallower spots, again the best formed reefs are mostly deeper where there is a healthy abundance of soft corals and larger sponges. Squirrelfish and soldierfish are both common, usually loitering near the darker recesses of the reef waiting for nightfall, and the reefscape is sometimes enlivened by the bright colors of porkfish.

The south coast of Jamaica is less visited by divers, although there are a number of reefs. The coastal shelf is wide here so there are no dramatic walls, but the distance from land means less pollution and sedimentation so there are some quite healthy coral populations. Fishing is still heavy, but on some reefs divers will be entranced by busy reef scenes dominated by bluehead wrasse, Spanish hogfish, chromis, and reef butterflyfish.

To the west of Kingston, the massive Portland Bight Protected Area provides a conservation model for the region. On land it includes large areas of hill and forest which are home to the Jamaican iguana, a giant lizard only found here, and to the coney, a small rodent (similar to the hutias of Cuba). At the water's edge there are extensive mangrove forests, and crocodiles are still common. Offshore, seagrasses

give way to wide areas of coral, while in the deeper water dolphins and even sperm whales have been seen. A lot of people live within the park, including fishermen, and the approach to park management has been to give local people the power to protect their own natural resources, and to train them as managers and wardens. So

Graysbys are one of the smaller groupers and are still found on most Jamaican reefs.

far this has proved far more successful than bringing in outside experts.

Other than fishers, few people visit the remote islands and coral reefs of Morant Cays to the southeast of Jamaica or Pedro Cays to the southwest. As a result, very little is known about these sites, but it seems likely that they include some very rich coral reefs surrounded by clear waters. Officially there has been some protection for these islands since 1907, but in reality the pressure for fishing has seen the establishment of small huts on several of the islands where fishermen stay overnight. Turtle nesting is also reported on a number of the smaller islands. Fishers also regularly visit the Grappler Bank and Formigas Reef which lie to the northeast of Jamaica, but again there is little or no information about the underwater communities.

HAITI

Temperature	23-30°C (Jan-Feb), 27-33°C (Jul)
Rainfall	Port-au-Prince: 1 320 mm (other parts are much drier); wetter Apr-May and Aug-Oct
Land area	27 156 km^2
Sea area	127 000 km^2
N° of islands	154 (10 large)
The reefs	450 km^2. Very little studied, but there are known to be beautiful coral gardens and some areas of dramatic reef scenery. Overfishing has removed the largest fish from most reefs, and sediments from the land have impacted some reefs.
Tourism	There is very little tourism: cruise ships take passengers to a single isolated bay on the north coast, and there are just a few other coastal hotels and two dive centers.
Conservation	There are no marine protected areas, and the few fishing regulations are largely ignored.

The beautiful mountainous land of Haiti takes its name from the Arawak Indians who called it Ayti or "Mountainous Land". It makes up the western third of the island which was named La Isla Española by Columbus in 1492 – today the island as a whole is still called Hispaniola. French influence began in 1664 when the French West India Company took over much of the western end of the island, and in 1697 it was formally ceded to France by Spain and renamed Saint-Domingue. It became France's most prosperous colony thanks to a massive slave labor force. In 1791, inspired by the French Revolution, the slaves rose up in rebellion. Thirteen years of fighting decimated the country, but in 1804 independence was granted. Sadly for Haiti, its history since independence has been one of continuing instability and today the country is the poorest in the Americas.

Although there are considerable areas of coral reefs around Haiti, remarkably little is known about them. There are certainly areas of very beautiful coral gardens and some dramatic walls, but most reefs have been affected by intensive and almost unregulated fishing. Forest clearance from the mountain slopes has added to these problems. Soil erosion has released sedi-

Sponges like this strawberry vase sponge lend a wealth of texture and color to Haitian reefs.

Fire corals, with their distinctive white-tipped branches, carry quite a sting.

ments which, swept into the sea, have smothered and killed some of the coral reefs closest to shore.

In contrast to these problems, Haiti suffers far less from many sources of pollution than other Caribbean countries – away from Port-au-Prince there is little industry, the country uses few agrochemicals, and even piped sewage systems are unusual, so most sewage pollution remains on land rather than reaching coastal waters. Away from the areas affected by sediments, the coral communities are quite healthy, and larger fish are still found in a few places.

A barrier reef runs along part of the northern shore, separated from the mainland by a channel 30 meters deep. Close to Cap-Haïtien this coast also houses a small settlement, Labadee, which is maintained for the regular arrival of cruise ships. Visitors can enjoy beautiful beaches and some snorkeling and diving, although this area is largely inaccessible to Haitians.

The reefs on the mainland coast and around the tiny island of Les Arcadins have been visited by a few scientists, and offer a beautiful array of marine life, including many tube sponges and the delicate and quite unusual pink vase sponge. Here there are also lots of deepwater sea fans in surprisingly shallow water. Small fish such as butterflyfish and bluehead wrasse are widespread, although there are few large species. Close to St Marc the reef drops off in a dramatic vertical wall, and the fringing reefs all around the Ile de la Gonâve also drop steeply into deep water.

Female princess parrotfish pass in front of an azure vase sponge.

There are many beautiful reefs along the south coast. A small tourist center on Ile à Vache takes visitors to extensive shallow coral gardens, and some deeper reefs at around 16-18 meters. Smaller fish are again abundant: damsels, butterflyfish, smaller groupers, and the bright fairy basslets. Larger fish such as barracuda and even tarpon are also found here. The coastline and adjacent areas offer wonderful opportunities to explore a wider area by boat, or for a hike through the surrounding forests and villages.

NAVASSA ISLAND, USA

Way off to the west of Haiti, in the direction of Jamaica, lies a tiny, uninhabited speck of an island which rises with dramatic cliffs out of the deep ocean. Navassa Island has been a US Protectorate since 1857, and is managed, from Puerto Rico, as a wildlife refuge. The surrounding waters plunge down to a shelf at about 23 meters' depth. Rubble and sand dominate the more level areas, but the steeper walls are smothered in corals, including more unusual species such as the smooth flower coral and maze coral. The underwater scenery is very dramatic, with a wealth of sheltering holes and caves for fish. Larger fish are common, including large groupers and snappers, and the bigger parrotfish. The offshore waters are also rich in the larger pelagic species, including hammerhead sharks and the occasional bull shark.

Most visitors to Navassa require permits, although Haitian fishers are allowed to come, and regularly make the 55 kilometer journey in tiny boats to fish these rich waters. At the present time their influence is considered to be relatively limited and the waters are still home to some of the healthiest coral reefs for many miles around.

Large snappers such as this dog snapper and cubera snapper (behind) are common around Navassa Island.

Haiti and the Dominican Republic (Hispaniola)

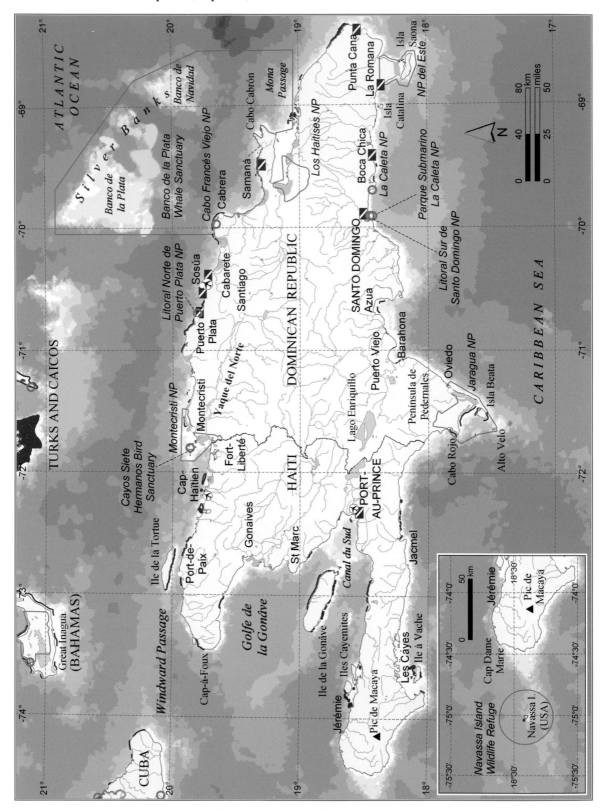

DOMINICAN REPUBLIC

Temperature	21-27°C (Dec-Feb), 24-30°C (Jun-Aug)
Rainfall	Santo Domingo: 1 380 mm; higher rainfall May-Oct; wetter in the north and east
Land area	48 444 km²
Sea area	261 000 km²
N° of islands	54 (5 large)
The reefs	610 km². There is a considerable variety, with some beautiful and unspoiled reefs, particularly in the national parks and on the Silver Banks. Close to the larger tourist centers many reefs are somewhat degraded.
Tourism	The second largest numbers of any Caribbean island, with 2.9 million international arrivals plus some 211 000 cruise ship passengers. There are 24 dive centers, mostly associated with the main tourist destinations, that take visitors to a broad array of sites on the coast.
Conservation	Many of the reefs, as well as adjacent areas of seagrass and mangrove, fall within national parks. There are growing efforts to improve the effectiveness of these parks, often encouraged by the dive centers which are keen to reduce levels of fishing and prevent further damage to the reefs.

The Dominican Republic is the second largest island nation in the Caribbean, making up two thirds of the land area of Hispaniola. Much of the country is dominated by hills and mountains, rising to Pico Duarte, the highest mountain of the Caribbean islands at 3 175 meters. Interestingly the country also claims the lowest land surface in the Caribbean – in the dry southwest of the country the surface of the inland salt lake, Lago Enriquillo, is actually 48 meters below sea level.

Columbus first came to these shores in 1492, and indeed he recorded the coral reefs in his ship's log. Both Arawak and Carib peoples were found on the island, but both groups were quickly decimated by European diseases and through fighting with the new arrivals. The Spaniards chose to build the first European city of the Americas in Santo Domingo, including the first cathedral, monastery, hospital, and, by 1538, even a university. Despite this early flourishing, Spanish attention quickly diminished as they turned to the even greater wealth they were finding in Central and South America. A checkered history, including two periods of US occupation in the 20th century, has now, at last, led to stability, and the country has been developing rapidly in recent years. Although agriculture is important, the extensive white sand beaches have become the island's major selling point and it is one of the Caribbean's major tourist destinations.

Although Spanish is the predominant language, English is widely spoken, partly as a response to the growing tourism from the USA, but also because many people have relatives there.

There are some truly spectacular reefs around the coastline of the Dominican Republic,

King crabs, like many crustaceans, are nocturnal – often seen in holes during the day, they come out and clamber over the reef at night in search of algae, their main food.

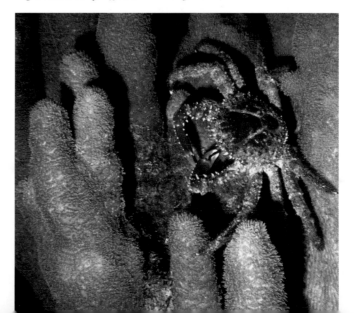

with a rich fauna of corals and fish. The Montecristi Barrier Reef is extensive, with healthy corals, large numbers of fish, and beautiful scenery. Far offshore, the Banco de la Plata also has wide expanses of coral reef, as well as offering some of the best opportunities for whale-watching anywhere in the world.

In many coastal areas, rapid development has degraded the reefs through overfishing, pollution, and sedimentation. There is a growing awareness of these problems, but much more needs to be done if these reefs are to recover.

ISLAND TOUR
The southwest

The southernmost part of the Dominican Republic, the Península de Pedernales, is one of the driest regions of the Caribbean, with near desert conditions in places. This is also one of the least developed areas of the country. The forests and savanna of the Jaragua National Park give way to fabulous, undeveloped beaches – smaller and more exposed bays with fringing reefs to the east, and long, sheltered beaches with high cliffs to the west. Many fabulous animals are found in Jaragua. Sea turtles nest on the beaches and there are large numbers of flamingos. The Jaragua lizard, the world's smallest land vertebrate, was discovered here in 1998 – it is fully grown at only 28 millimeters. In stark contrast the rhinoceros iguana, which also has a major stronghold in this part of the island, is one of the largest Caribbean lizards – males have been recorded to 120 centimeters and weighing some 10 kilos.

There are fringing reefs around Cabo Rojo, with elkhorn and fire corals in the shallower waters, then a wall dropping from 18 to 45 meters with abundant star corals and sheet corals. Isla Beata is surrounded by reefs, with some seagrasses on the western side. Stony corals are again abundant, including brain, star, and finger corals. Further offshore, some corals are found on the steep, stony flanks of the small volcanic island of Alto Velo, although there are no true reefs.

Most access to the far southwest of the country is from Barahona where there are now a few large hotels. There are fringing reefs nearby, with well developed spur and groove formations.

HUMPBACK WHALES

Every year some 10 000 humpback whales migrate from the icy waters of the Arctic to spend their winter in the warm waters of the Caribbean. Up to 6 000 of these come to the Silver Banks. In the Arctic waters they feed continuously on the rich supply of small fish and krill (a large shrimp-like crustacean). Heavy with fat reserves they make the journey of several thousand kilometers to breed and produce their calves. Amazingly they do not usually eat when they are in the warmer winter grounds.

Whale-watching trips are run between December and April, and this is one of the few places in the world where it is possible (under supervision) to swim and snorkel with these magnificent beasts. The regulations allow for so-called "soft encounters" – the whales cannot be chased, but swimmers are placed in the water and the whales often approach. Typically it is the females with calves who tend to move more slowly and so are likely to be observed – the females regularly reach 15 meters in length and their "little" calves are typically 4 or 5 meters long.

One fairly common activity on these breeding grounds is breaching, when these huge whales jump fully clear of the water like vast torpedoes, crashing back with a tremendous splash. It is likely that breaching is a form of display or signaling (males tend to breach more often than females), but it may also help to dislodge parasites. Some have suggested it could even be a form of play. Other activities include slapping the water with a flipper or a tail, and spy-hopping, where the whales lift their heads and upper bodies out of the water, then fall back. It is even possible to hear the humpbacks sing – males have some of the most complex songs in the animal kingdom, and a single song with themes and repeated phrases can last for 30 minutes. All the males on the bank will use the same song, which evolves and changes from year to year.

Larger groupers and even nurse sharks are sometimes seen in the deeper water.

Across the mouth of Puerto Viejo a string of mangrove islets marks the top edge of a small barrier reef. The sheltered lagoon is carpeted with seagrass and occasional small patches of corals, while on the outer edge there are some areas of elkhorn coral, with boulder brain coral descending into the deeper water.

The southeast

Most of the accessible areas of coral reef on the south coast lie to the east of Santo Domingo. Many divers from the large tourist center of Bahía de Andrés visit the Parque Submarino La Caleta. Here there is a fringing reef which slopes gradually down with high spur and groove structures from 10-18 meters then a dramatic drop to 25-40 meters. There is a great diversity of coral, and even more sponges, including branching tube sponges, brilliant red rope sponges, and black ball sponges, with giant barrel sponges in the deeper water. Lobsters and king crabs are often seen on the reef slope.

The park has a number of wrecks including the 38 meter *Hickory*, and these are home to numerous fish. Sergeant majors tend to collect in large numbers on the outside of these wrecks, while dense schools of sweepers, deep red soldierfish, and the occasional moray eel are often seen within. In the sandy areas around the base of the reefs and the wrecks it is quite common to see garden eels. These long slender eels live in burrows in the sand – they are extremely shy and withdraw when approached, leaving an apparently bare field of sand. Watched from a distance, however, they are mesmerizing as they rise slowly out of the sand with their heads held at 90 degrees like flags. They waft gently in the water and occasionally pick at tiny pieces of zooplankton.

Across the mouth of Bahía de Andrés lies a long reef, known as Boca Chica. The deeper parts of this reef are healthy, with areas of staghorn coral and some tall towers of pillar coral. Sheet corals, star corals, and boulder brain corals become more common at around 30 meters. Black durgon are quite often seen in groups swimming above the reef – although apparently dull black from a distance, a closer view reveals beautiful, subtle shades of blue, orange, and yellow,

particularly around their faces. Yellowhead wrasse are often seen here, usually alone, looking for food in the nooks and crannies of the reef. These close relatives of the bluehead wrasse have strange sleeping habits. In the hour before sunset they can be seen in small groups loitering about over small patches of sand. If you wait long enough you will see one dive down then wriggle head first into the sand. The others will follow shortly after. They remain buried like this – totally invisible – until the morning.

Further to the east there are fringing reefs, generally in poor condition, along the coast towards Isla Catalina. Towards the north of this island there is a dramatic wall quite close to the shore. In shallow water, pretty, and healthy, coral communities harbor countless colorful fish such as chromis, parrotfish, and butterflyfish. The wall drops almost vertically down to a sandy base at 40 meters, and is richly encrusted with sponges

A few healthy stands of elkhorn corals in the shallows of the National Park del Este provide important shelter for grunts and schoolmasters.

and sheet corals, with bushy black corals and slender whip corals on the steepest parts.

The National Park del Este covers a large area of land and sea in the far southeast. The western and southern sides of the mainland and Isla Saona have wide sandy beaches and gently shelving offshore waters, while the eastern sides and the northern shores of Isla Saona are rocky.

The back reefs behind the Montecristi Barrier Reef are populated with an impressive array of corals, including branching and blade fire corals, finger corals, and staghorn corals.

The park includes important areas of seagrass and mangroves, with manatees in shallower areas, while dolphins occasionally come up to boats. The best developed reefs are on the western side, where wide gardens are dominated by the tall wafting branches of sea plumes, with some colonies of elkhorn and staghorn coral in shallower water. Large southern stingrays are regularly seen here.

Along the southern shore of Saona strong currents offer opportunities for drift diving. Deeper on the reef slope it is quite common to see nurse sharks resting on the bottom. A well formed fringing reef has also developed between Isla Saona and the mainland at the eastern end of the channel close to the tiny island of Catalinita. Here, facing the ocean, divers are sometimes treated to views of eagle rays, manta rays, and even hammerhead sharks.

The eastern tip of the Dominican Republic has major hotel developments along its extensive sandy beaches, particularly around Punta Cana. From this point northwards to Punta Macao stretches a long barrier reef, teeming with life on its outer slopes, although parts of the lagoon close to shore have become quite polluted. Towards the south this reef has dramatic walls, with barracuda sometimes hovering in the waters offshore, and lobsters and king crabs hiding in the darker crevices. Farther north the profile of the reef becomes more gentle, with a greater range of corals including brain corals, cactus corals, and areas of elkhorn coral. The inquisitive and gregarious yellowtail snapper is common in places, and smaller fish are abundant everywhere.

The north coast

The mountainous peninsula of Samaná has a scattering of reefs off its northern shore. Those towards the western end do not have large amounts of stony coral, but there are many smaller fish and also schools of Bermuda chub which feed on the algae that predominate here. Further east the reefs are healthier, and small, sheltered coves support vigorously growing fields of elkhorn coral. At the tip of the northern Cabo Cabrón the coastline is breathtaking, with 400 meter cliffs plunging into the sea and continuing underwater in a wall falling to depths of 150 meters or more. Hard corals smother the rocky surfaces, dominated by star corals, brain corals, and sheet corals, while the remaining spaces are carpeted with a colorful palate of sponges. Humpback whales occasionally come very close to shore in this area during the winter months. Dolphins are more regular visitors, and divers will routinely see Spanish mackerel or cero, and perhaps even reef sharks.

Rapid and poorly planned hotel development between Cabarete and Puerto Plata has had devastating effects on the coral reefs closest to shore, especially near Puerto Plata. Here the mangrove forests that once protected the reefs from sediments and nutrients washing off the land have been replaced by hotels and golf courses which produce large amounts of pollution, and 80 percent of the coral has died. Things are a little better around Sosúa and there are still extensive soft coral gardens with patches of elkhorn, fire, and brain corals. French and queen angelfish add flashes of brilliance to the dramatic scenery deeper on the reef slope. Like most angelfish these species eat sponges, and have developed powerful digestive systems to deal with the toxins they contain. The spotted drum is also seen here – adults are often wary, but juveniles, with their fantastic flag-like fins, are regularly observed, quite unafraid, at the base of corals or in small crevices.

The far northwest of the country, towards Haiti, is bounded by the Montecristi Barrier Reef. Large areas of mangrove forest line the coast and there are extensive seagrass meadows just offshore. In the deep lagoon coral patches are populated by gorgonians, sea rods, and also star corals, and some areas of finger and staghorn corals. The "back reef" on the land side of the barrier reef has some fabulous gardens, with extensive fields of quite slender branching finger coral, staghorn, and fire corals. Smaller fish – including dense schools of grunts and snappers – tend to be more common in these lagoon and back-reef areas, with many spending their juvenile phases in these waters before moving offshore as adults. Elkhorn coral used to dominate in the shallow, wave-swept waters on the outer reef edges, and there are still a few areas where these are growing, although in many places they have been replaced by fire coral. Deeper down, star corals, with some brain corals, have formed massive colonies which may be centuries old. Complex scenery has grown up with great pillars, deep caves, and crevices. Many ships have been wrecked in this region over the years, including three Spanish ships which sank in 1593. Only one was salvaged at the time and the search is on for the remaining cargoes, which included more than 300 000 silver coins, as well as gold and jewels.

Silver Banks

Far offshore to the north of Samaná are two large, shallow banks – Banco de la Plata and Banco de Navidad, commonly referred to in English as Silver Bank and Christmas Bank, or simply the Silver Banks. Geologically speaking, these are the last two in the long chain of shallow banks which make up the archipelago of the Bahamas and the Turks and Caicos Islands. These banks have no islands, but parts of the reef become exposed and dry at low tide. There are coral reefs along the northern edge of Banco de la Plata and the northeastern side of Banco de

Navidad. In the shallow water there are still some areas of elkhorn corals, while deeper down the topography of the reef becomes very complex with high towers of coral rock, richly decorated with stony corals and sponges. The northern edge drops rapidly to very deep water.

Unfortunately these banks have been heavily fished, and the few people who have snorkeled or dived in the area have reported damaged reefs with few large fish. Indeed the low numbers of fish, conch, and lobster have led to

Cero are sleek ocean-going members of the tuna family, and regularly come in over the reefs.

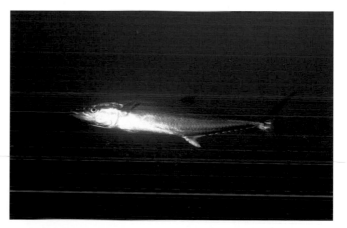

Steel gray forms of Bermuda chub typically swim high above the reef but will sometimes be seen grazing on algae.

reductions in the numbers of fishers heading to the banks in recent years. Despite this, the banks attract several hundred tourists per year, in the winter months, to see the humpback whales (see page 142).

Puerto Rico

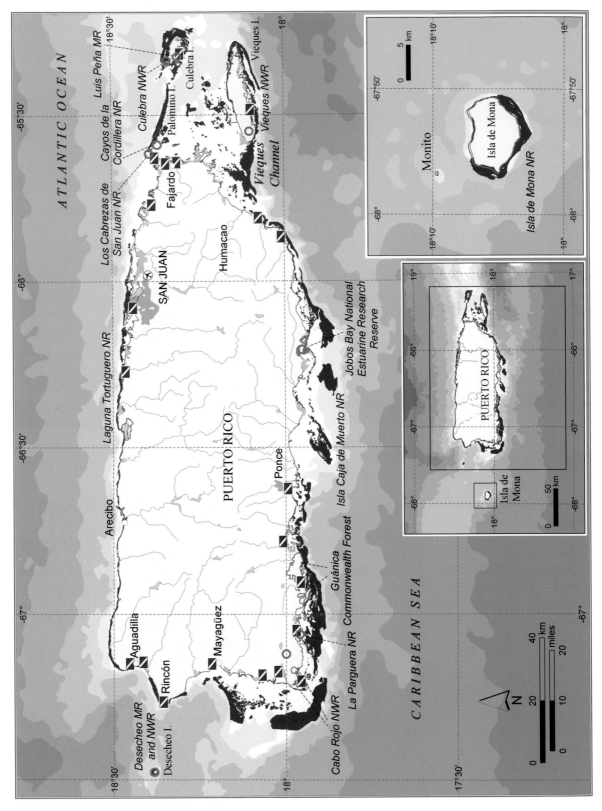

PUERTO RICO

Temperature	21-28°C (Jan-Feb), 24-31°C (Jun-Sep)
Rainfall	San Juan: 1 340 mm; considerably drier on the south coast; wettest May-Nov
Land area	9 063 km²
Sea area	205 000 km²
N° of islands	152 (12 large)
The reefs	480 km². The island is surrounded by coral reefs, with the most vigorous growth being extensive coral gardens around the offshore islands and dramatic walls along much of the south coast and the western islands of Desecheo and Mona. Around the mainland large areas of the reefs are degraded and there are almost no large fish.
Tourism	The largest numbers of any Caribbean island, with 3.6 million international arrivals plus some 1.4 million cruise ship passengers. There are 32 dive centers and this is a rapidly growing sector. Tourists come to most areas, but particularly the east and southeast.
Conservation	There is now a lively interest in coral reef conservation. Although there are only a few marine reserves, more will be designated which should give the reefs a chance to recover from years of overfishing.

The island of Puerto Rico is a Commonwealth of the USA – not a fully integrated state, but something quite close. It is an island where the Hispanic and North American cultures are uniquely blended. Columbus arrived in Puerto Rico in 1493 – his large expeditionary force only spent two days here, and named it San Juan Bautista, before moving on to Hispaniola. In 1508 Juan Ponce de León came to explore the island and founded the town of Caparra on the excellent natural harbor which was named Puerto Rico. Over time the town took on the name of the country (San Juan) and the country took on the name of the harbor, Puerto Rico. The island slowly developed an agricultural economy, largely based on sugar and coffee. It remained a Spanish island until the Spanish-American War of 1898, at which time it was ceded to the USA.

Much of the island has the feel of a North American state, with US stores and companies dominating, and English a widely spoken second language. The smaller villages still have a strong Hispanic feel, and Hispanic architecture dominates Old San Juan, where some of the forts date back to the 16th century, and Ponce (pronounced PonSay), where many buildings date from the boom in sugar and rum production between 1890 and 1930.

Much of Puerto Rico is hilly or even mountainous, while to the north the waters sink down into the deepest point in the Atlantic Ocean, the Puerto Rico Trench, some 8 605 meters below the sea surface. There is considerable urban development along the coast, and

Spotted morays can reach a meter in length, but often only their heads are seen peering out from dark recesses.

The smallest of the Caribbean butterflyfish, the longsnout butterflyfish is often found below a depth of 15 meters. It is a shy species and is usually solitary.

fast highways connect most parts of the island. Inland there are still a few patches of forest, such as El Yunque rainforest to the southeast of San Juan, and the Guánica Commonwealth Forest – a tropical dry forest. For the more adventurous Puerto Ricans and travelers these offer spectacular hiking, and a view of Caribbean islands which is now increasingly rare.

Puerto Rico is surrounded by coral reefs, and diving and snorkeling are rapidly growing in popularity amongst locals and visitors. In recent years many areas of coral reef have suffered from a combination of hurricanes (in 1998 Hurricane Georges was particularly damaging), coral diseases, and chronic overfishing. Many reefs have been degraded or, like Jamaica, taken

The lettuce sea slug grazes algae, but in a sleight of digestive ability, keeps the photosynthesizing parts of the algae undigested. These not only provide them with their green camouflage, but continue to convert sunlight into sugars as an extra source of food.

over by algae. Despite this, beautiful areas remain – dramatic walls and cave scenery are found along the south coast, with more gentle gardens to the east. Some of the healthiest reefs are found around the offshore islands.

Increasing efforts are now underway to protect the marine environment. There are a few marine protected areas, but many more are urgently needed to help not only the coral reefs but also the struggling local fishing industry.

ISLAND TOUR

Most of the accessible diving in Puerto Rico is in the east. In the northeast the scattering of tiny islands, La Cordillera, are very popular, with dazzling white beaches that descend into gently sloping coral gardens dominated by soft corals. Slightly deeper, the sand patches provide a shimmering turquoise backdrop to the pretty reef scenes, and in a few places there are deep crevices, overhangs, and tunnels. Stingrays are sometimes seen here, and eagle rays occasionally fly through, but these are also great places to look close up at the corals and invertebrate life. The Pederson cleaner shrimp is often found nestling amongst the protective tentacles of a sea anemone. These beautiful shrimps, with long white tentacles and translucent bodies marked with fine white lines, perform a critical role in cleaning parasites and dead skin from other fish (see Chapter 1.1). They can sometimes be persuaded to clean people – the diver must make a gentle approach, then slowly proffer an open hand. Occasionally they will flit across and proceed to move gently over the skin looking for edible scraps, almost too light to feel.

There are extensive reef areas dominated by soft corals in the Vieques Channel. The southern side of these reefs marks the shelf edge and the beginning of the dramatic reef wall. Deepwater sea fans are common on the wall, while schooling jacks and Bermuda chub often swim past.

Deep water comes in close to the island all along the south coast. Typically the reef falls gradually at first, with corals growing on gentle slopes, and with well developed spur and groove structures out to the top of the wall, at a depth of around 17-25 meters. From here the bottom falls away precipitously. The water is usually very clear, thanks to the proximity of deep water, and to there being fewer sediments coming off the land on this, the dry side of the island.

Above the wall, the landscape is dominated by sea rods and sea plumes, with occasional pillar corals. Green morays are often seen lurking in this forest of corals. Rope sponges, whip corals, and encrusting lettuce corals are found on the steeper sides of the spurs, and there is also a tremendous variety of sea anemones. The corkscrew anemone has delicate translucent tentacles, decorated with faint traces of white in corkscrew-like twists, while the strange branching anemone has short "pseudo-tentacles" which emerge from narrow crevices in the rock – they actually have much longer tentacles which they extend at night. The giant anemone is perhaps the most beautiful, however, with long creamy white tentacles each with an enlarged tip in pink or purple. On the steep wall, sponges abound, but here there are also encrusting and brain corals, and occasional black corals. Black durgons are common in the waters just off the wall, and Atlantic spadefish are also regularly seen.

Along the west coast, erosion has sculpted some beautiful scenery with deep cuts and caves in the old reef rock. Larger fish are not common, but the surprising colors of the fairy basslet light up the reef, and sponges abound. The northern coast is almost continuously pounded by heavy seas, and so is rarely visited by divers. It has a scattering of fringing reef structures, but for the most part these consist of only a few live corals, together with algae and sponges, growing on ancient limestone rocks.

Offshore islands

Puerto Rico's four larger offshore islands are surrounded by some of the best coral reefs in the country. To the east, both Culebra and Vieques have small settlements; both have been used as military bases, and indeed Vieques still has an important military presence. In the past, parts of these islands were bombed during training exercises – this has damaged some areas of reef (and there may still be some unexploded ordinance in places); however the military presence has also protected the reefs from the impacts of development.

Culebra, with its scattered islands and rocks, is surrounded by coral reefs, giving a spectacular backdrop of turquoise and indigo to its bright, uncrowded beaches. There are wide coral gardens suitable for snorkeling as well as diving, but there are also deeper reefs, with walls on some of the northern shores. Soft corals tend to dominate, but brain and star corals are widespread, with occasional pillar corals rising up above their surroundings. Pillar corals are one

BIOLUMINESCENCE

One of the most extraordinary sights on a night dive or snorkel is to turn off all the flashlights and watch the fireworks. Countless organisms in the plankton produce light when disturbed, and in the inky black waters (particularly on a moonless night) the movement of a hand in the water can leave a trail of fabulous blue-green sparks.

Among the most abundant producers of this light are tiny creatures called dinoflagellates. Just a fraction of a millimeter across and invisible to the naked eye, they are capable of giving off a powerful flash of light through a simple chemical reaction. Mosquito Bay, on Vieques, and Phosphorescent Bay at La Parguera, have some of the greatest bioluminescence in the Caribbean. Here the numbers of dinoflagellates in the water can reach hundreds of thousands in every liter. Boats passing through the water carve wide swathes of brilliance, but it is while swimming that the effects can be best appreciated – the greater the antics of the diver or snorkeler the more impressive the display, with clouds of sparks echoing every movement.

There are various reasons why an organism might benefit from being able to emit light, but these tiny dinoflagellates are particularly unusual: for them the flash of light acts as a burglar alarm. When small predators of the dinoflagellates, such as copepods (a crustacean), are sensed, the dinoflagellates give off a flash of light. This draws in fish that eat the copepods (but not dinoflagellates, which are just too small to be of interest) – a blunt but effective system of justice.

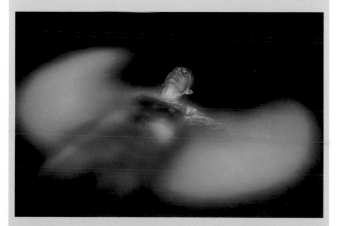

of the few stony corals to be active during the day: up close the surface is smothered in fine tentacles which the corals use to filter potential food items from the water (corals generally only use this form of feeding as a top-up to their diet). The northern shores are home to a few groupers – coneys and graysbys, but occasionally also tiger, yellowfin, and Nassau groupers. Between Culebra and the small island of Luis Peña, a marine reserve was established in 1999, where all fishing is forbidden. There are already more fish here than anywhere else, and numbers will increase in the coming years. Some of the small islands, as well as the northwest peninsula of Culebra itself, are home to colonies of hundreds, even thousands of brown boobies and noddies.

The eastern half of Vieques is still a military base. Reefs surround the island, but are best developed to the south, where they face towards deeper water. There are no shelf-edge walls accessible to divers, but there are smaller drops descending to areas of sand and rubble. Nurse sharks are quite often seen resting on the bottom,

A rather slender pillar coral rises up in deeper water.

sometimes under overhangs. These big sharks tend to be active from dusk to dawn, when they patrol the reef for food – their diet is almost entirely invertebrates such as lobsters and mollusks, but they occasionally take small fish.

The two islands to the west are uninhabited, and are surrounded by crystal clear, clean water. Dive boats and day trippers sometimes go to Desecheo, but trips to Mona require more planning, with many visitors choosing to camp on the island. Both islands are protected, and the waters around Desecheo have been closed to all fishing since 2000.

Desecheo has the healthiest and densest cover of stony corals in all Puerto Rico. Here the full variety of Caribbean corals can be admired, with wide areas of pillar corals, brain corals, boulder star corals, and a variety of solitary cup corals in the darker recesses. There are areas with steep walls and caves, and fish life is abundant almost everywhere. Smaller groupers and parrotfish are regularly seen, and occasionally larger species such as rainbow and midnight parrotfish. If the protection of these waters can be properly enforced it is likely that more and more large fish, including jacks and snapper, will be seen here in the coming years. In deeper water the longsnout butterflyfish is quite abundant. These delicate beauties are often solitary, and are quite shy. Their slender mouths are used to pick at a range of foods, including the feeding filters of Christmas tree worms.

On land, Mona is dry, with some forest and shrubs. It is home to the huge Mona ground iguana (related to the rhinoceros iguana of the Dominican Republic) and also a small snake, the Mona boa. Mona is encircled by gently shelving slopes with extensive areas of healthy corals, including brain and star corals, with some wide areas of elkhorn and the symmetrical brain coral. Further offshore there is a shelf break and the vertical walls plunge into the deep ocean, populated by black corals, whip corals, and encrusting and sheet corals. Groupers are regularly seen, and horse-eye jacks often swim past, but these islands are perhaps best known for their turtles – it has been estimated that about 1 000 hawksbill turtles live around Mona and its adjacent Monita, feeding on the diverse life, including sponges, tunicates, and small crustaceans and mollusks.

US Virgin Islands

Temperature	22-28°C (Jan-Feb), 26-31°C (Jun-Sep)
Rainfall	1 060 mm; wetter Aug-Nov
Land area	350 km²
Sea area	6 000 km²
N° of islands	82 (8 large)
The reefs	200 km². Most of the reefs are patch and fringing reefs close to the islands, with other areas of coral-encrusted rocky scenery. A barrier reef marks part of St Croix, with steep walls on the north shore.
Tourism	592 000 visitors, with a further 1 891 000 cruise ship passengers. Diving is popular, with 27 dive centers on three islands.
Conservation	Many years of decline are now being addressed with genuine concern. Large marine parks are beginning to offer real protection, and turtle nesting sites in St Croix are set aside. Benefits are already being detected, with greater numbers of large fish in some sites.

The Virgin Islands lie like a scattering of pearls at the end of the long chain of the Greater Antilles. Close to Puerto Rico, they are separated from the Lesser Antilles by the 2 000 meter deep Anegada Passage – the deepest channel connecting the Atlantic Ocean to the Caribbean Sea.

The islands to the south and west of this archipelago are officially described as an "unincorporated territory" of the USA – largely self governing, but the people are US citizens. These islands were purchased from Denmark in 1917 largely because of their strategic importance – the Anegada Passage was a major shipping lane. They are hilly islands, mostly dominated by scrub and forest, but each is in fact highly distinctive. St Thomas, the tourist center, is the most developed island, St Croix is the biggest and still retains much of its old Danish character, while St John is small and remains largely undeveloped.

Today the US Virgin Islands are actively promoting marine conservation. Marine parks dominate the seascape and many areas are already showing signs of recovery after years of decline. Away from these areas overfishing remains a problem, while occasional failures of sewage systems and outputs from the vast numbers of pleasure boats create localized pollution. The rum factory on St Croix has created an 8 kilometer (5 mile) dead zone through its release of warm and toxic effluents. Sediment run-off from the land adds to the problems and the high incidence of coral diseases in the Virgin Islands may be symptomatic of the stresses caused by these combined threats.

Thanks to the considerable efforts of the diving industry, mooring buoys have greatly reduced anchor damage at the dive sites. A broader shift in attitudes towards the marine environment began in the late 1990s, since when protected areas have been expanded and new sites have been designated. Fishing has been banned from several sites, including the entire Virgin Islands National Park. The US Virgin Islands now have some of the largest marine protected areas in the region.

ISLAND TOUR
St Thomas and St John

The long, hilly island of St Thomas is highly developed, especially around the capital Charlotte Amalie and towards the east. By contrast St John is largely undeveloped, with most of the population concentrated on the west coast around Cruz Bay. Over half of St John lies within the Virgin Islands National Park, first established through the vision and generosity of Lawrence Rockefeller who, in 1956, gave some

The Virgin Islands

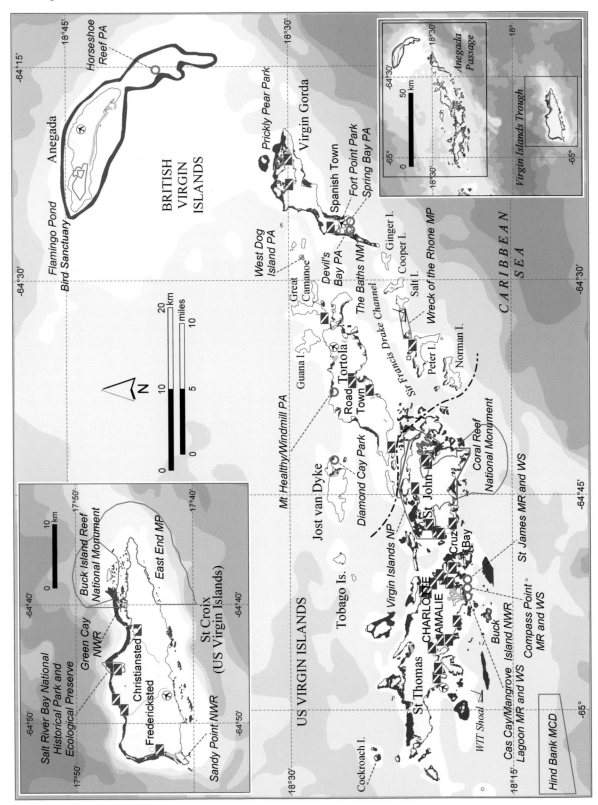

2 000 hectares to the US government. It has since been expanded and this is now one of the largest areas of undeveloped land in the small islands of the Caribbean. The park is criss-crossed with hiking trails through forest and scrub, many leading to unspoiled beaches.

Fringing and patch reefs are found around both islands as well as around the numerous smaller islands scattered across the adjacent sea. Most reefs are some distance offshore, but a few, particularly off St John, are accessible to snorkelers and divers from the shore. There are quite a number of wrecks too, including the WIT *Shoal*, a 100 meter wreck with five deck levels that sank here in 1984 and is now smothered in a great profusion of life.

Many of the popular diving locations lie in the waters between St Thomas and St John (Pillsbury Sound). Soft corals abound on typically undulating scenery, while in a few places there are more dramatic caves and canyons. Smaller fish, including tangs, hamlets, and angelfish are abundant, and eagle rays and tarpon occasionally visit the reefs around the small islands in the north of the channel. The Virgin Islands are a stronghold for the blue bell tunicate, one of the most eye catching of this abundant class of animals also known as sea squirts. They live in small clusters on rocky surfaces where they filter the water for food, rather like sponges, drawing it in through a hole or siphon, and then discharging it through a smaller hole. It is the young tunicates which have attracted the interest of science however – quite unlike the adults, these are free-swimming larvae, shaped a little like a tadpole. These larvae have a notochord, a sort of primitive spinal chord, which in evolutionary terms is seen as the precursor to the spine – and these simple sea squirts are actually more closely related to fish, even to birds, reptiles, and mammals, than any other animals on the reef.

Since its establishment, the Virgin Islands National Park has been expanded considerably to include a large marine area with extensive coral reefs, seagrasses, and mangrove forests. Immediately adjacent to the Park, the Virgin Island Coral Reef National Monument was declared in 2001. Fishing is now banned from most of these areas and, although a small amount of illegal fishing continues, fish num-

bers are likely to increase considerably in the coming years.

Offshore to the south and west of St Thomas there are coral patches, pinnacles, and areas of dramatic rocky scenery encrusted with corals and sponges. Larger fish are more abundant, including a variety of jacks. These silvery fish are active predators, usually traveling in small schools or packs. They feed on other

Red hind are popular with fishermen. To protect fish stocks, one of their main spawning grounds in the US Virgin Islands has now been placed strictly off-limits.

Some beautiful stands of elkhorn coral are to be found in shallow water in the Buck Island Reef National Monument.

fish, and can occasionally be seen making a concerted attack. The activity is a frenzied blur as the jacks make repeated dives into a school of smaller fish, which swirl about in confusion, or crowd down amongst the corals for cover. Bar jacks are common in many sites, while horse-eye jacks are often seen on deeper dives or near to wrecks. The rainbow runner, a more slender species, is usually found in areas of open water. One other species sometimes seen here is the deep bodied permit – which usually feeds on crustaceans, mollusks, even sea urchins.

Further south, the Hind Bank Marine Conservation District is closed to all fishing and anchoring year round. Red hind travel considerable distances to spawn at this site once a year. During the first 12 years of protection at the site, the average size of the red hinds increased from 29 to 39 centimeters. Such increases in length make a truly disproportionate difference to a fish's reproductive output. A 39 centimeter fish will be about two and a half times the weight of a 29 centimeter fish, but will produce eight times as many eggs. Protecting spawning sites such as this helps fish stocks to recover over very wide areas indeed.

St Croix

St Croix lies some distance from the rest of the Virgin Islands, and is physically separated by the 4 500 meter deep Virgin Islands Trough.

GHOST FISHING

One of the most widespread forms of fishing in the Caribbean islands is the use of fish traps. These are thrown into areas of shallow water on or near coral reefs and marked with a small buoy. Fish enter the traps, usually tempted by bait, but cannot find their way out.

In the past such traps were made from natural materials, including wood or palm fronds, tied with natural fiber rope, and if a trap was lost it would quickly degrade and fall apart. More recently, traps have been made from wire mesh. When these are lost they continue "ghost fishing" for months or years before the wire mesh corrodes – fish are still caught, but nobody makes use of the catch.

There is a simple solution to avoid this waste. Biodegradable panels built into the walls of traps mean that, if lost, they will only continue to catch fish for a few weeks at most. Such panels are a requirement by law in a number of countries including the US Virgin Islands, but enforcement is difficult until the fishers can be made to see that it is in their own interest to comply. As with so many environmental issues, legislation needs to go hand in hand with education.

Underwater, it presents a quite different world from the other islands. The north coast faces very deep water, and spur and groove formations lead down in many places to vertical walls populated with plate corals and the dark bushy forms of black corals. Some of the most dramatic scenery is to be found in the area of Salt River Bay, where a natural fault has created deep walls running in towards the island. Heading east from Christiansted to Buck Island the reef becomes a barrier reef, separated from land by a wide lagoon.

The Buck Island Reef National Monument (not to be confused with the National Wildlife Refuge off St John) includes over 7 000 hectares of shallow water, entirely closed to fishing and anchoring. The site offers some great snorkeling, with high coral cover and areas of elkhorn and other hard corals even in quite shallow water. This site lies adjacent to the East End Marine Park, established in 2003. Between them, these sites protect a vast area of coral reefs and open sea – fishing is prohibited over wide areas, and the already healthy coral reef communities are likely to flourish in coming years. Further east still, shallow water continues in an area known as the Lang Bank. Little is known about the sea floor here, but there are certainly small reef patches and also areas of seagrass and gorgonian communities.

Turtles are abundant in the waters around the Virgin Islands, and St Croix has some of the most important nesting beaches. In the east the Jack and Isaac Bays are a critical nesting ground for green and hawksbill turtles, while the extensive beaches of the Sandy Point National Wildlife Refuge in the southwest of St Croix are the most important nesting site in the northern Caribbean for the leatherback turtle – the world's largest turtle. These giants are recorded at over 2 meters long and weighing 500 kilos, and are known to travel for thousands of kilometers across the Atlantic Ocean, regularly swimming as far north as Canada during summer months. Volunteers working at the site gather important information about the turtles and also help in the relocation of nest sites when eggs are laid in areas of beach erosion. As a result of these protection efforts the number of turtles coming back to nest at the site each year appears to be slowly increasing.

BRITISH VIRGIN ISLANDS

Temperature	22-28°C (Jan-Mar), 25-31°C (Jul-Aug)
Rainfall	996 mm; wetter Aug-Nov
Land area	161 km²
Sea area	81 000 km²
N° of islands	52 (15 large)
The reefs	330 km². Mostly patch and fringing reefs near the islands, with further rocky coral outcrops. Extensive platform reefs around Anegada.
Tourism	296 000 visitors, with a further 203 000 cruise ship passengers. There are 11 dive centers on two islands. Most tourism is concentrated in Virgin Gorda and Tortola, but all islands are popular and this is a world center for yacht-based tourism.
Conservation	Good protection around moored dive sites and the three main marine protected areas, but the islands urgently need to develop a more complete plan for marine conservation.

The British Virgin Islands (see map, page 152) make up the northern and eastern parts of the Virgin Island archipelago, at the eastern end of the Greater Antilles chain. Christopher Columbus arrived here in 1493 and named the islands Las Once Mil Virgenes, after "the 11 000 virgins", disciples of St Ursula, presumably inspired by the sheer number of islands that scatter the waters of this archipelago. Today the islands, usually simply known as the BVI, are a British Overseas Territory.

Most of the islands are hilly and covered with scrub or patches of forest. Large parts of the coastline are rocky, but there are many miles of white sand beaches, particularly on the northern shores. The second largest island, Anegada, which is some distance to the northeast, is low lying, almost flat, and formed from limestone.

There are extensive coral reefs around most of the islands, with fringing reefs close to the shore and smaller patch reefs scattered across the shallow shelves. Generally, the sea floor slopes gently so there are no dramatic walls, but dive operators have found numerous scenic locations with pinnacles and canyons, arches and small caves. Horseshoe Reef is the largest reef system, encircling Anegada and extending far down beyond the southeast tip of the island. In the more sheltered bays there are patches of seagrass, and several small mangrove forests.

The BVI is a great place for seeing turtles, particularly green turtles. These are often unafraid of divers, offering an excellent opportunity to approach and admire them. Sadly, on beaches where there were once hundreds, perhaps thousands of nesting turtles, now only a handful come to nest each year.

The BVI was probably the first country in the world to have mooring buoys installed at popular dive sites in order to prevent anchor damage. In a program led by the diving industry, some 200 buoys have been installed since the early 1980s, and today the corals around the dive sites are among the least damaged in the country. Unfortunately anchor damage remains widespread elsewhere.

Male sailfin blennies flare up their impressive dorsal fin in part of a complex and dramatic series of displays to attract the attention of females and to fend off rival males.

Overfishing is a problem, largely because of tourist demands and high levels of recreational fishing. There are, however, good numbers of fish around the dive sites, and, because of fears about ciguatera poisoning, some fishers avoid the larger predatory fishes, so these are also sometimes seen. Further offshore quite a number of recreational game fishers come over from the US Virgin Islands, as marlin, barracuda, wahoo, and bonitos are more abundant in the BVI. A small number of protected areas have been established to protect coral reefs, but some fishing is still permitted even in these areas.

Tourism has added to the pressures on the marine environment. Since the 1970s there has

Sergeant majors are abundant in shallow waters, here around a pillar coral. They have a highly varied diet but often feed on plankton.

been widespread coastal development, and the islands also now have the highest concentrations of charter yachts in the world. Yachts cause considerable damage to corals and seagrasses with their anchors, and also release tons of sewage directly into nearshore waters. Efforts are being made to reduce these impacts through education, regulation, and the provision of increased numbers of mooring buoys. Many of the older hotels and other coastal developments also add to sewage pollution, while the clearance of mangroves and other forests has exacerbated problems of sedimentation in some areas.

Turtle fishing is still permitted outside the nesting season. The turtles are caught in nets and their meat is sold in local supermarkets, and large numbers of eggs are also dug up from the nests. This turtle capture is a source of concern not only in the BVI, but also in the adjoining US Virgin Islands where great efforts are being made to protect turtles. If you see turtle meat for sale in a restaurant make a fuss and refuse to eat there.

ISLAND TOUR

As most of the sites that divers visit are protected with mooring buoys, there is remarkably little damage to the coral at the main dive sites. There has also been very little fishing here, making it possible to see quite large fish, and in many places dense schools of grunts which just hang in the water and seem to drift with the surge. The British Virgin Islands is also one of the few places in the world where tourists can be taken to see coral spawning. This spectacular event has only come to the attention of scientists in the last two decades, but in the British Virgin Islands the timing is sufficiently well known for divers to be taken to watch it – one week after the first full moon in August each year.

The main island, where three quarters of the population live, is Tortola, while the chain of islands to the southeast, across what is known as the Sir Francis Drake Channel, is highly popular with divers. Many of these islands are rich in bird life and are protected as bird sanctuaries. Boats often crowd the waters all around, but beneath the surface there is a rich profusion of life. In shallower water the bright striped sergeant majors crowd around visitors, and yellowtail snapper are abundant. Deeper down among the coral scenery there are many angelfish, and this area also has some very large hogfish, with males reaching a meter in length. This species is all too frequently overfished in other parts of the Caribbean. Tarpon are quite common in this area, and nurse sharks are regularly seen resting in caves on deeper reefs. Another remarkable fish – no less spectacular but requiring a whole different way of looking – is the sailfin blenny, which reaches only about 5 centimeters in length, and inhabits small holes in areas of rocky rubble and sand. The males are dark brown to black, while females tend to be paler. But their distinguishing feature is a high, sail-like dorsal fin which they raise up in display. In the early morning and late afternoon males can be observed "jumping" out of their holes, flaring their high fins to impress nearby females. They are also highly aggressive towards one another – females will chase off other females and males have been observed in quite vicious mouth-to-mouth combat. If you see a little black head peering out of the sand it is well worth moving a short distance away and waiting for some action.

Undoubtedly the best known dive destination in this area is the wreck of the RMS *Rhone*, a Royal Mail steamship built of steel and 100 meters (310 feet) in length. She sank here during a hurricane in 1887 with only a handful of survivors. The ship lies in two parts on the ocean floor close to Salt Island. Visits to a wreck of this age are fascinating, as the entire surface is encrusted with over 100 years of life: sponges, sea fans, tunicates, colonial fanworms, plume corals, and stony corals. Countless fish have made their home here too, including dense schools of snapper, and French and bluestriped grunt. Barracuda are also often seen. This site, including the island of Dead Chest, is a marine park – although a small amount of fishing is permitted under license, in reality there is very little anywhere near the dive sites.

From Tortola divers also visit reefs and areas of dramatic rocky landscapes to the northwest, particularly around the islands of Jost van Dyke and Tobago. In a few places there are spectacular towers of rock, with caves and canyons inhabited by glassy sweepers. Sponges tend to dominate on the rocky surfaces, and in deeper water jacks and rainbow runner sometimes swim past.

Diving is also popular off Virgin Gorda, particularly around the rocky islands known as The Dogs. Here there is some highly varied scenery, with rocks and short walls encrusted with sponges, as well as more gently sloping gardens with extensive areas of sea rods and sea plumes. In a few places there are dense schools of tomtates, and, against the steeper walls and rocky areas, blackbar soldierfish, squirrelfish, and glasseye snappers are often seen alone or in small schools waiting for nightfall. The southwestern tip of the island includes The Baths Natural Monument and the Spring Bay Protected Area. This is a beautiful area of sculpted granite boulders on the beach and continuing into the

sea – the sheltered bright waters are excellent places to introduce newcomers to snorkeling. There are large numbers of smaller fish all around, and a few pretty corals.

The remote and barely populated island of Anegada is now, once again, home to the Caribbean flamingo. There were thousands of these wonderful birds in the British Virgin Islands in the 19th century, but when their young became a popular food item they were effectively exterminated. Since 1995 small numbers have been reintroduced, and a growing breeding colony has now become re-established on Anegada. The island itself is encircled by coral, with a total reef area greater than all the other islands combined. These reefs have healthy coral populations dominated by star and mustard hill corals. There is an abundance of fish, with a greater number of large predators such as groupers and snappers than on most other coral reefs in the Virgin Islands. Anegada is also famous for its shipwrecks – over the years hundreds of ships have foundered here. The reefs extend in a long curve over 20 kilometers to the southeast in the Horseshoe Reef Protected Area. This area is closed to most visitors and fishing is restricted to local Anegadans, who take fish and over 15 tons of lobster per year (although it sounds a lot, over such a large area this may be sustainable). Little studied, these may be some of the most important reefs in the country.

EXCLUSIVE CONSERVATION

The British Virgin Islands has a number of privately owned islands. These include a few very special places where concerned owners have made the environment a key selling point for exclusive tourism. Three in particular stand out: Mosquito, Necker, and Guana Islands. Each caters for low-volume tourism, with just a few chalets set amongst largely natural surroundings. On each island efforts have been made to remove grazing animals such as goats, allowing beautiful forests to grow up and rare native species to flourish. Rainwater is collected as the main source of water, and no sewage enters the surrounding sea.

Guests, as well as enthusiastic scientists, are rewarded with a taste of a wild Caribbean, without cars or crowds. On Necker and Guana they can even see some of the rarest creatures of the region such as the red-legged tortoise, rock iguanas, and even Caribbean flamingos. With unpolluted waters all around, and often with lower levels of fishing, these islands also offer fabulous opportunities to see the adjacent marine communities.

Lesser Antilles

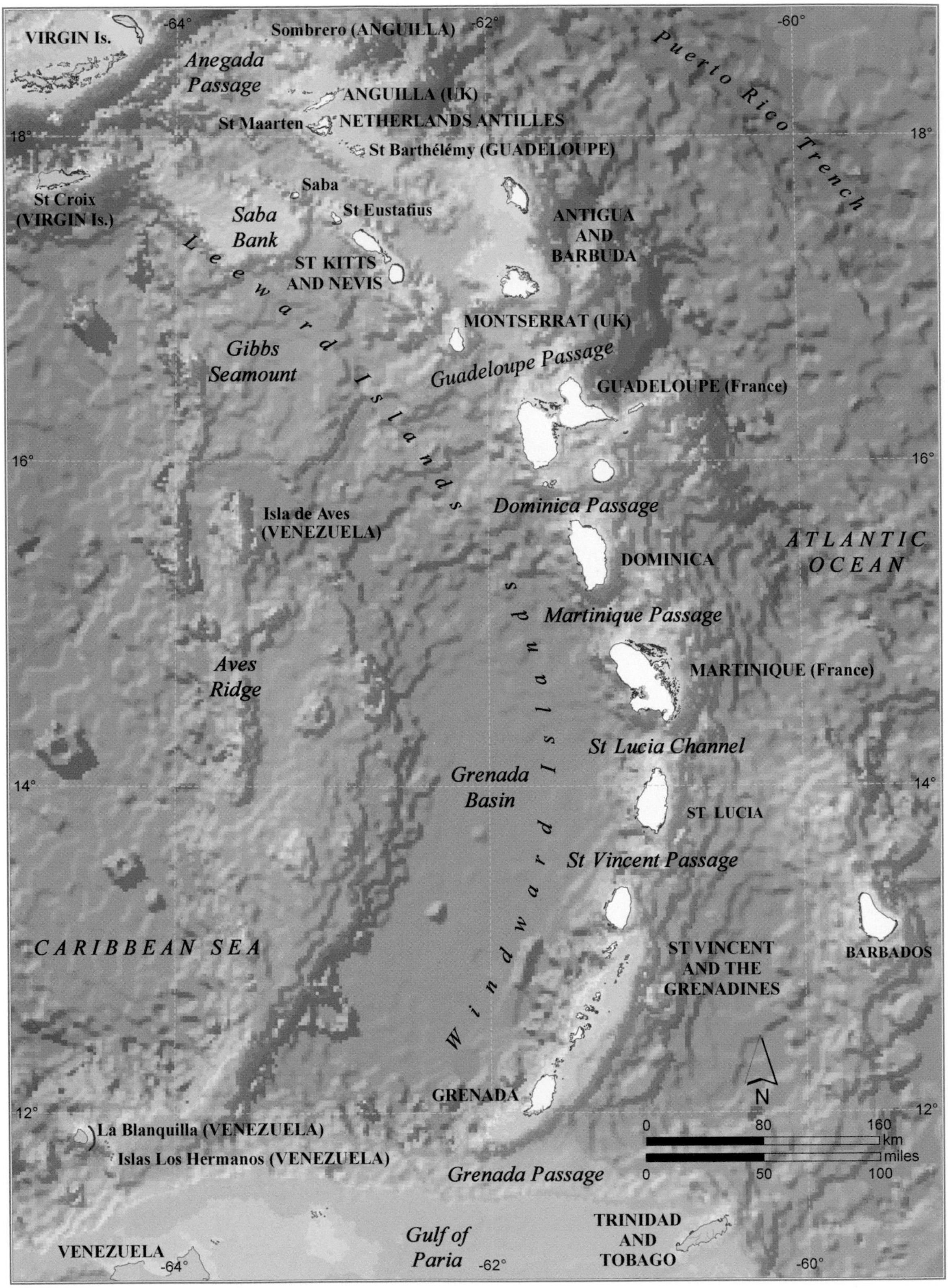

VIRGIN Is.   -64°   Sombrero (ANGUILLA)   -62°   -60°

Puerto Rico Trench

Anegada Passage

ANGUILLA (UK)

St Maarten   NETHERLANDS ANTILLES

18°   St Barthélémy (GUADELOUPE)   18°

Saba

St Croix (VIRGIN Is.)   *Saba Bank*   St Eustatius   ANTIGUA AND BARBUDA

Leeward Islands

ST KITTS AND NEVIS

MONTSERRAT (UK)

Gibbs Seamount   *Guadeloupe Passage*   GUADELOUPE (France)

16°   16°

Isla de Aves (VENEZUELA)   *Dominica Passage*   *ATLANTIC OCEAN*

DOMINICA

Martinique Passage

Aves Ridge   MARTINIQUE (France)

St Lucia Channel

14°   *Grenada Basin*   14°

ST LUCIA

St Vincent Passage

Windward Islands

CARIBBEAN SEA   ST VINCENT AND THE GRENADINES   BARBADOS

GRENADA   N

12°   12°
La Blanquilla (VENEZUELA)   0   80   160
Islas Los Hermanos (VENEZUELA)   km
                                        miles
Grenada Passage   0   50   100

TRINIDAD AND TOBAGO

VENEZUELA   *Gulf of Paria*   -62°   -60°
-64°

2.4 LESSER ANTILLES

This chain of tiny islands stretches along the eastern edge of the Caribbean Sea. It includes a staggering diversity of landscapes and seascapes, filled with both natural splendors and fascinating cultures. Rugged volcanoes simmer and steam, their steep slopes cloaked in fabulous forests. Great Atlantic waves pound the eastern shores, while bright beaches nestle amongst the hills to the west. Underwater is an unexpected and diverse array of life – coral reefs are most widespread around the non-volcanic islands such as Anguilla, Antigua and Barbuda, and Barbados, while the geologically young coastlines are dominated by volcanic walls, fissures, and vast boulders. Corals have begun to populate these scenes too, but bare surfaces remain, home to such oddities as seahorses and frogfish.

All of these islands have their origins in the movement of the Caribbean plate – a large section of the Earth's crust – towards the western Atlantic. In a slow motion collision which has been going on for more than 70 million years, the Caribbean plate has pushed itself over the Atlantic sea floor. At the same time the edge of the plate has crumpled, throwing up mountains and volcanoes – the islands and mountains of the Lesser Antilles. Today there are at least ten active volcanoes. Montserrat has been erupting since 1996, while the others are a little quieter, but offer fascinating opportunities to see volcanoes up close, with wide areas of desolate landscapes, crater lakes, and pools of boiling mud.

The climate is fairly constant all year round, and a little drier in the north. A continual warm breeze is provided by the northeast trade winds which dominate the climate for up to 300 days a year. These winds greatly helped the trade to these islands for many centuries, and the islands are sub-divided into the Windward Islands in the south, where the sailing boats from Europe first arrived (Grenada to Martinique), and the Leeward Islands in the north.

Despite the dramatic backdrop of the volcanoes, it is the tranquil beauty of these islands which has attracted most outside interest. Extensive white sand beaches, combined with warm, welcoming cultures and friendly people, have led these islands from small agricultural nations to world-class centers of international tourism. Many are attracted by the underwater scenery, and come here specifically for the diving, while increasing numbers have a first snorkel or dive while visiting. Perhaps even more important, however, is that the local people are also beginning to take an interest in the marine environment, and are taking up diving and snorkeling as a form of recreation.

Fishing remains an important activity, with fishers using traps and seine nets, as well as hooks and lines, and sometimes spears, to capture reef fish and lobsters. This has led to reduced fish stocks over wide areas – larger fish or lobsters are rare and in some places even parrotfish and small groupers are hard to find.

The way forward is clear however. Exciting developments in the Netherlands Antilles and St Lucia vividly show how properly managed protected areas can benefit everyone. Working together, fishers, the tourist industry, and conservationists have established some impressive marine reserves. Fish and corals are flourishing, local people are gaining further employment in the growing diving industry, and fish catches are increasing as fish spill out from the marine reserves into the surrounding areas.

The longlure frogfish is often found over the volcanic scenery of the Lesser Antilles. A master color changer, this individual has adopted all the hues, and imperfections, of a pink sponge.

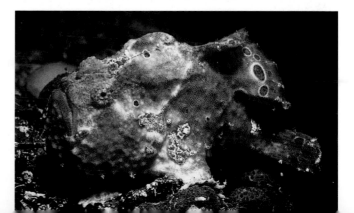

ANGUILLA

Temperature	26°C (Jan-Mar), 29°C (Jul-Sep)
Rainfall	1 020 mm; peak Aug-Dec
Land area	86 km^2
Sea area	90 000 km^2
N° of islands	23 (15 very small)
The reefs	Less than 50 km^2. The reefs are largely shallow patch and fringing reefs offering beautiful coral gardens, rarely extending below 30 meters.
Tourism	48 000 visitors; no cruise ship arrivals. Four dive centers. Many tourists come for the beaches. Most of the diving is along the north shore where there are fringing and patch reefs and a number of wrecks.
Conservation	Five marine parks have been established. There is some overfishing so larger fish are not common.

The tiny island of Anguilla (see map, page 163) lies at the northern end of the Lesser Antilles. A territory of the UK, Anguilla has a strong sense of independence. Everything is small scale on this small island – there is a small fishing industry and even an offshore finance industry. The island is heavily reliant on tourism but, because of its size, even this is small scale, and the island is famed for friendly people and beautiful beaches.

Anguilla is relatively flat, rising to only 65 meters, and with mostly shallow surrounding waters. Besides the main island there is a scattering of small offshore islands, including the remote Sombrero Island, little more than a rock with a lighthouse.

Corals are found all around the island in fringing and patch reef communities. The best developed are probably those off the north coast in a system running parallel to the land but some distance offshore. The scenery of these reefs is generally not dramatic – there are a few steep walls but, because Anguilla sits on a wide shelf, the deepest edges of the reefs are typically only

Trumpetfish are often seen lurking amongst the dense foliage of sea plumes, waiting in ambush for smaller fish to eat.

20-30 meters below the surface. Despite this, diving and snorkeling around Anguilla can be very beautiful. Wide coral gardens are dominated in most places by extensive fields of wafting sea plumes, sea fans, and sea rods, as well as areas of pillar corals, brain corals, and star corals. There are occasional schools of blue tangs, and chromis are abundant. Moray eels are also often seen here; if you come across one out in the open, watch from a sensible distance. If it is hunting you might see it poking its head into the nooks and crannies of the reef looking for fish or crustaceans.

There are healthy if inaccessible coral communities around Dog Island to the west of Anguilla, while the waters around Sombrero Island to the north – though little known – are also thought to have some important coral communities. Until quite recently there was a somewhat surprising plan to develop a satellite launching pad on Sombrero, but environmentalists campaigned hard against it and won. Although the island is small, it is home to an important seabird nesting colony as well as a small black ground lizard found nowhere else in the world.

Quite a number of ships have been sunk to make popular dive sites. The largest wreck, at 70 meters (230 feet) in length, is the *Sarah*, which sank during Hurricane Klaus in 1984 and was towed to her present resting place the same year. After more than 20 years underwater this boat is encrusted with life, including soft corals, sponges, solitary cup corals, and even a large number of Atlantic thorny oysters – a vibrant patchwork of color completely masking the bare steel of the boat. The oysters are often hard to spot, as their sculpted spiny shells are themselves highly encrusted with life. Look out for a narrow, dark line between the two "valves", which slam shut as soon as the oysters sense your presence.

The Anguillans have established a number of marine parks where no anchoring is permitted, but marine conservation measures are otherwise somewhat limited. Fishing is quite intense, and it seems likely that many species are overfished. For the most part the reefs are populated by small fish and the occasional barracuda. The largest fish populations are probably around Scrub Island, and bull sharks and blacktip sharks occasionally come to this area.

Brown chromis are abundant, swimming high above the sea bed picking at microscopic food from the plankton.

WHERE DID THE BEACH GO?

In small islands undergoing rapid development one critical raw material is sand – not just for tourists to lie on but for builders to make cement and building blocks. With few alternatives, sand is often taken from beaches or dunes. Sand mining is internationally widespread, but taking sand from beaches is almost invariably ill advised – many such beaches undergo rapid erosion, especially during hurricanes. On the tiny island of Anguilla extensive dunes up to 6 meters high were mined from Sile Bay in the 1980s. This was okay until Hurricane Luis hit the island in 1995 and seawater swept some 46 meters inland. Recognizing this problem, some countries have switched to dredging large volumes of sand from the ocean floor. But this too has its problems. Although sand may look quite stable – both on beaches and underwater – it is constantly shifting with the movements of waves and currents. Many beaches are constantly supplied with new sand swept in from offshore, so even offshore sand mining can have quite unpredictable consequences.

Far from land these problems may be less acute. In the Bahamas large-scale sand mining has taken place in remote shallow bank areas, largely for export to the USA where it is used not only for building but to replenish eroding beaches. Even in the Bahamas, however, there are problems. Close to Nassau, the removal of sand from beaches for the building of golf courses has led to rapid erosion.

Efforts to undo the damage of sand mining are expensive, involving importing sand and dumping it on the beach, or building large walls or other buffers to attempt to reduce erosion. In many cases such efforts have failed, and have even led to increased erosion.

Perhaps the best solution, as adopted in the Netherlands Antilles, is to obtain sand from far inland. Here sand may be mined directly, or even "produced" by crushing limestone rocks.

NETHERLANDS ANTILLES
SABA, ST EUSTATIUS, AND ST MAARTEN

Temperature	Saba: 24-27°C (Feb), 28-34°C (Jun)
Rainfall	775 mm; no clear wet/dry season
Land area	68 km^2
N° of islands	9
The reefs	Dramatic scenery abounds in many places, particularly in the volcanic surroundings of Saba and St Eustatius, where numerous and often quite large fish, as well as turtles, are plentiful. St Maarten is best known for its wrecks.
Tourism	421 200 visitors (of which 402 000 go to St Maarten), plus 867 800 cruise ship passengers. There are 12 dive centers, with three each on Saba and St Eustatius and six on St Maarten.
Conservation	Some of the leading efforts in coral reef conservation are to be found on these islands, and the results are healthy or recovering reef systems.

The Netherlands administers a small scattering of islands in the Caribbean; the northern group is described here. Although these islands lie in the Leeward Islands of the Lesser Antilles they are sometimes referred to as the Windward Netherlands Antilles, so as to distinguish them from their sister islands of Bonaire and Curaçao which lie off the coast of Venezuela (see page 216). These others lie "downwind" and are thus referred to as the Leeward Netherlands Antilles.

The Windward Netherlands Antilles consist of two and a half islands – Saba, St Eustatius, and the southern half of St Maarten (St Martin). Two of the islands, Saba and St Eustatius, are renowned for their dramatic underwater scenery with coral communities growing on dark volcanic rock, while in St Maarten reefs are more extensive. In each of the islands there is strong and growing concern for the natural environment and their surrounding waters are probably some of the best protected in the region.

ISLAND TOUR
Saba

The tiny island of Saba, an old and long-dormant volcano which last erupted in 1636, rises precipitously from the ocean. Any visit starts with a breathtaking landing on one of the region's shortest runways, built on almost the only patch of flat land halfway up the volcano slopes. The island has one road (The Road) which winds tortuously through the four villages, including the capital known as The Bottom. It is an island for divers and nature lovers, and there is no beach tourism simply because there are no accessible beaches.

All around the island to a depth of 60 meters is the Saba Marine Park, effectively set up in 1987, although not officially declared until 1998. This park encompasses some of the best

Many Caribbean islands are home to unique species, such as the beautiful Saban anole which is found only on Saba.

diving in the Lesser Antilles. Dive tourism generates significant income for the local people, but one of the great joys of Saba is the small scale of everything. There are only three dive operators and no large hotels. The park is run by the Saba Conservation Foundation, which employs a number of staff and maintains some 40 mooring buoys. The local people fish in most areas, but their total catch is quite modest, and some of the best reefs are closed to all fishing. In other words the reefs are largely unimpacted.

Underwater the scenery is unusual. Steep cliffs come to a halt in quite shallow water, and a shelf surrounds most of the island. Apart from a small area in the south around Giles Quarter, the corals have not built a true reef. Rather the scenery consists of massive volcanic boulders and high towers of dark rock, all encrusted with corals, sponges, and sea fans. Between these are wide valleys of volcanic gray sand. Some distance off the west shore is a group of three pinnacles rising up out of very deep water. Diving on these is spectacular, and there is every chance of a visit by large species such as eagle rays, manta rays, and even a bull shark or a Caribbean reef shark. Turtles are commonly seen around the island, even mating pairs. Here too is a great place to witness the groupers and snappers that have all but disappeared from many other reefs in the region.

Anguilla and the Netherlands Antilles

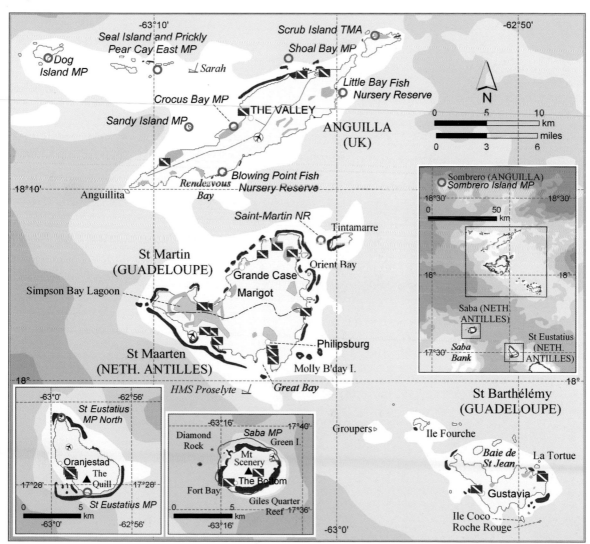

The giant volcanic boulders around Statia are thickly encrusted with corals, sponges, and other life, here giving shelter to a queen angelfish.

Inaccessible to most people, there are more reefs to the west of Saba on the wide Saba Bank. This area has been known to fishermen for years, and may be a "drowned" coral atoll (its shallowest points are still more than 10 meters below the sea surface). There is reported to be good coral cover on parts of the bank, but it has received very little attention from divers or scientists.

No visit to Saba is complete without a hike to the top – Mt Scenery – although at over 800 meters it is no easy walk. At the summit the

constant swirling of cloud has created a cloud forest where low trees are totally smothered in mosses and ferns. Be sure to take advice from the dive centers before doing this hike, and avoid it after a deep dive in the morning – it is high enough to affect the gases dissolved in your bloodstream, with the risk of getting a bend.

St Eustatius

This small island, often known as "Statia", has only recently joined the tourist itinerary and still receives relatively few visitors. Its small size, steep topography, and dark sand beaches have protected the island from mass tourism. Like Saba it is a volcanic island, with steep shores all around and a dramatic, though dormant, volcano known as The Quill at its southern end.

There are no clearly formed fringing reefs close to the island, but corals are abundant on the volcanic rock and in a few places have built up walls, and spur and groove formations. Much of the spectacular underwater scenery is provided by basalt rocks and lava flows which have left long tongues of rock, or huge blocks and boulders, often fragmented in quite regular, grid-like formations. Soft and hard corals as well as sponges are abundant. There are some fabulous barrel sponges, as big as divers – such sponges may only grow 1-2 centimeters a year, so the big ones can be more than 100 years old.

Marine conservation is taken seriously, and in 1996 a marine park was declared in the surrounding waters to a depth of 30 meters. Active management began in 1998, run by the St Eustatius National Parks (Stenapa). Two large sectors of this park have been declared marine reserves where no fishing is permitted, and these encompass some of the best diving areas. To the south there is an impressive drop-off with gorges and canyons. Offshore from the main town of Oranjestad are a number of historic wreck sites. Although the wrecks themselves are no longer visible, occasional artefacts can still be found.

For divers, Statia presents a relatively new and certainly an exciting destination. Visitors should take the time to look around the island, which boomed as a free trading port from the mid to late 1700s; there are some beautiful old colonial buildings from this period. There are also extensive walking trails around the Quill National Park.

MARINE PARKS – DUTCH STYLE

In the waters around Saba and Statia (as well as Bonaire in the Leeward Netherlands Antilles), a whole new approach to looking after marine life has been established. All of the waters around these islands have been declared marine parks and are managed by an independent agency. There is a comprehensive plan to protect marine life for each island, developed in collaboration with all the different users, and applied sensibly to avoid conflicts.

Within these parks some of the most important areas have been set aside as reserves where fishing is strictly prohibited, and large numbers of mooring buoys have been put down so that there are no boat anchors to damage sensitive areas.

The parks also charge a modest fee – typically US$5-10 for single use or US$10-30 for annual use – to divers and visiting yachts. Income from this does not fully fund the management of the parks, but it is a critical contribution and helps to emphasize the value of the reefs to the island governments. Most visitors are happy to pay this fee, and indeed many feel it should be higher, as they are glad that their money is being used to protect some of the Caribbean's best diving.

St Maarten

Tourism boomed earlier at St Maarten, but now at last the environment is beginning to receive some interest and attention. Many visitors come here for luxury hotels, beaches, and gambling, and only a small proportion visit the reefs, mostly on one-off visits. The island does not have a reputation as a popular diving destination but this may be slowly changing. Although there are problems of pollution, and visibility may not be as good as around other islands, there are still things to see. Many dive operators also make day trips to the adjacent islands, including Saba and Statia.

Most of the reefs are to be found to the south. Although there is a little coastal snorkeling to be done off some of the smaller bays, the real reef starts 2-3 kilometers offshore. There are quite a few reef patches due south of Great Bay, but probably the least disturbed reefs are found to the east, including those close to the cluster of small islands of Molly B'day, Pelican Rock, Hen and Chicks, and Cow and Calf. These reefs are home to a wealth of small fish, including the lurking shapes of nocturnal species such as squirrelfish and soldierfish, waiting in the shadows for nightfall. Small wrasses and damselfish abound, and schools of blue tang often swim past. The small islands are important for seabirds, so keep an eye out for pelicans, tropicbirds, and brown boobies. Another popular dive site in this area is the 200-year-old wreck of the British frigate HMS *Proselyte*, complete with cannons and large anchors, now totally encrusted with reef life. More recently, in the 1990s, a considerable number of wrecks were sunk to provide further dive sites.

It is also worth pointing out the wide Simpson Bay Lagoon that makes up a large part of the west of the island. Although there are no reefs here, there are some interesting mangrove forests and important areas of seagrass. It is quite likely that some of the fish seen on the reefs began their lives as juveniles in these mangroves, or may come here to breed.

Considerable efforts led by the Nature Foundation St Maarten are being made to protect the island's reefs. A comprehensive marine park has been proposed, covering most of the nearshore waters, with two strict nature reserves encompassing the most important reef areas. In addition about 25 mooring buoys have been placed to reduce anchor damage. With the extra protection that these measures should provide, it is likely that the reefs will become home to increasing numbers of fish.

A Spanish hogfish picks at crustaceans and bluehead wrasse look for scraps on the highly grazed surface of a newly sunk wreck.

Yellow tube sponges lie below some of the rugged volcanic scenery for which Saba and Statia are famed.

St Kitts and Nevis

Temperature	25°C (Jan-Feb), 28°C (Jun-Oct)
Rainfall	1 170 mm; peak Jul-Nov
Land area	275 km^2
Sea area	10 000 km^2
N° of islands	3 (1 small)
The reefs	180 km^2. Fringing and patch reefs are widespread, and particularly well developed in the waters between the two islands. Volcanic rock and lava provide alternative scenery to the south of Nevis.
Tourism	70 600 visitors, plus 252 000 cruise ship arrivals. Four dive centers. The majority visit St Kitts, but substantial numbers go to Nevis and many visit both islands.
Conservation	There are some healthy reefs, and turtles are often seen. No significant conservation efforts in the marine environment.

St Kitts and Nevis is the smallest wholly independent nation in the western hemisphere. The islands were originally named by Columbus, but these names have since evolved. St Kitts is still, in official texts, known as St Christopher (Columbus' patron saint), while Nevis may once have been Nuestra Señora de las Nieves (Our Lady of the Snows), perhaps

St Kitts and Nevis

because the summit was perpetually wreathed in white cloud like snow.

The islands have a bloody history. The original Carib inhabitants struggled to repel British and French forces, but in a massacre in 1629 more than 2 000 Caribs were killed by the British at Bloody Point on St Kitts. This was followed by sporadic fighting between French and British troops, and the islands changed hands several times until eventually becoming a British colony in 1783. The sugar cane established at that time is still an important crop today, although, of course, tourism is now the major industry.

Both islands are dominated by steep volcanic mountains. The 1 315 meter high Mt Liamuiga on St Kitts offers exciting hiking through rich rainforest to a central crater and steam vents. Volcanic activity also shaped much of the underwater scenery and there are some beautiful and exciting dives to be had over dramatic lava flows, particularly to the south of Nevis. In a few places to the north of St Kitts there are also "seeps" where water heated by the volcano rises up through the sand.

Coral reefs are scattered all around both islands, and there are also beautiful coral patches on other areas of shallow shelf. In amongst the corals crustaceans are abundant, including crabs, lobsters, and slipper lobsters. Green, hawksbill, and loggerhead turtles are frequent visitors, though few nest on the islands.

The beautiful southeast peninsula on St Kitts has been declared a protected area on land, and a marine site has been proposed around Sandy Point. But no marine protected areas have actually been set up to date, and overfishing has diminished the numbers of fish in many areas. There is a proposal to stop the capture of all turtles, but some are still taken in the open season (October to February).

ISLAND TOUR

The most extensive coral reefs are probably those along the eastern shore of St Kitts, although this side can be difficult to access as it is often swept by Atlantic swell. Patch reefs along the west coast are dominated by a mix of soft and hard corals,

The slipper or Spanish lobster has modified a pair of antennae into broad shovels which it uses to dig in the sand for invertebrate food.

A single colony of staghorn coral, less than a meter across, is probably only a few years old. Such colonies can grow to form wide fields.

but sponges also abound, lending considerable color to the reefscape. The almost iridescent azure vase sponge is always a special sight, while tube sponges can be yellow, orange, or purple, and there are red rope sponges as well as the bushy branching form of the green finger sponge.

Identifying sponges can be very difficult, for while they appear to have distinctive shapes and

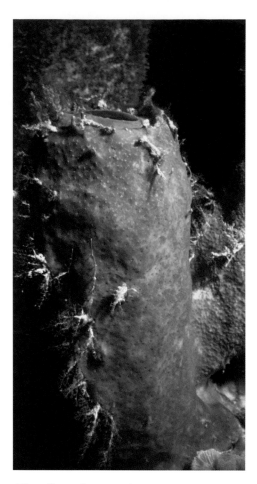

The yellow tube sponge is common on many reefs. Small gobies sometimes take up residence inside the tubes.

hues, some species can be any of several colors and their shapes are unreliable indicators. In some cases scientists can only identify them by taking a sample back to the lab and looking at the fine "spicules" which make up the sponge's skeleton.

Sponges are unable to move, so many have evolved complex chemical defenses to protect themselves from predators. A few creatures, however, seem unperturbed by these, and sponges are the preferred food of angelfish and the hawksbill turtle.

There are reported to be some 400 shipwrecks around the two islands, but few have been mapped and in many cases the only evidence of a shipwreck is a pile of stones. These were used as ballast – dead weight to hold a light or empty vessel lower in the water and to keep it stable.

Much of the tourist diving takes place on the western side of the shelf that runs between the two islands, where there is an array of shallow reefs with slightly deeper dives further from land. Here there are true reef structures, with wide areas of branching elkhorn and staghorn in the shallower water, giving way to sea fans, sea rods, and plume corals in deeper water. Schools of creole wrasse are often seen towards the deeper water – these blue-purple plankton feeders almost always collect in large schools. Like many reef fish they change sex with age: individuals up to about 15 centimeters long are female, then as they increase in size become male, and are characterized by longer fins and a blotchy pattern with a few dark scales developing on a yellow background.

Diving at night around the islands it is common to see the strange armored form of the slipper lobster (also known as Spanish lobster). These unusual creatures are extremely shy, but can still be observed from a distance as they forage about on the reef. Evolution has worked hard on them, and the long antennae common to lobsters have been transformed into short, paddle-like shovels on the front of their heads. Look closely, however, and you can see a second pair of very short antennae (antennules) with a forked tip. These are packed with microscopic hairs that smell (or taste) the water for food. The lobsters constantly flick these antennules around in order to get a clearer smell and can even determine the direction from which a scent may be coming. They eat a variety of invertebrates, and sometimes use their shovel-like antennae to dig into the sand.

Well developed reefs are found along the west side of Nevis as well as out to the northeast. Towards the south are patch reefs as well as volcanic scenery with dramatic rocks and dark tunnels. Lobsters and king crabs are common, and squirrelfish and soldierfish often lurk in the shadows waiting for night time when they go out to feed.

Snorkeling can be done from the shore in a few places, notably from the bays on the southern end of the peninsula in St Kitts, and from Newcastle Bay and Pinney's Beach in Nevis. In general, however, visibility improves further from the shore, so it is often good to go out in a boat to a snorkeling site.

ANTIGUA AND BARBUDA

Temperature	22-28°C (Feb), 25-31°C (Aug)
Rainfall	1 052 mm; no clear peak: drier Dec-Apr
Land area	462 km²
Sea area	110 000 km²
N° of islands	19 (7 small)
The reefs	240 km². Both islands have extensive and well developed coral reefs, as well as some very interesting diving on rocky substrates. Recent hurricanes have damaged many of the shallow water corals.
Tourism	222 000 visitors, plus 409 000 cruise ship arrivals. Ten dive centers (nine on Antigua). Almost all the visitors go to Antigua, which is famed for its fabulous and secluded beaches.
Conservation	With such extensive coral reefs there are some very healthy areas with good fish populations. Although there are several protected areas, the actual protection these provide is minimal.

At first glance the two islands of Antigua and Barbuda have little in common. Antigua, gently hilly with a long twisting coastline guarding isolated beaches and sheltered anchorages, is the political and population center, with a large and important tourism industry. Barbuda, 40 kilometers to the north, is a low-lying island with a tiny population. Fishing is the main industry and few tourists visit.

Both islands are quite dry. They were once forested, but were cleared during the 17th century to make way for sugar cane plantations. By 1970, with the collapse of the sugar industry, the plantations were largely abandoned and have now been overtaken by scrub, with low thorny trees and cacti.

The offshore waters of both islands have extensive coral reefs, including fringing reefs and, particularly to the north and south of Antigua, barrier reef systems. There are also large areas where corals grow on bare rock. The greatest diversity of corals is to be found on the reefs further offshore where the water is generally clearer.

Like their neighboring islands, Antigua and Barbuda were hard hit by hurricanes in the late 1990s. Corals on the shallower reefs were severely affected, and have since become domi-nated by algae in many areas.

The fishing industry, which generally uses traps, has grown in recent years. In Antigua, fish, lobster, and conch are taken for local (and tourist) consumption, while Barbuda exports a considerable amount – particularly of spiny lobsters – to neighboring islands. The increasing number of fishers appears to have placed a strain on the ecosystems – catches have diminished and, because fishers generally target larger individuals, the average size of the remaining fish has decreased, particularly off Antigua.

Despite these pressures, an abundance of marine life is still to be found around these islands: many areas are in good health and offer exciting diving. The value of these resources has not, however, been fully recognized, and existing

The red-billed tropicbird is spectacular with its long tail plumes. These birds only rarely come in to land, when nesting in burrows on rocky offshore islands.

Antigua and Barbuda, and Montserrat

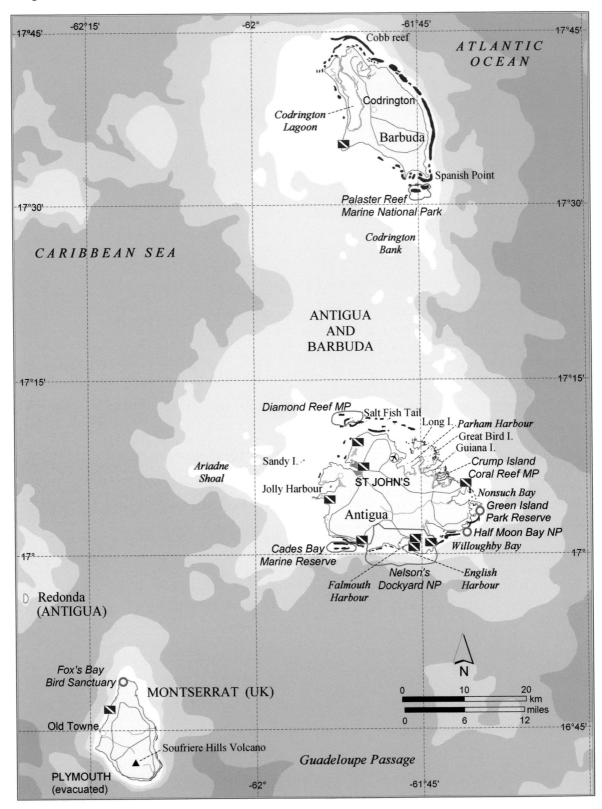

protected areas are poorly managed, if at all. Tourism development continues, but lacks the benefits of centralized planning. Guiana Island, just off the northeast of Antigua, is one of the last wild places in the country and highly important for seabirds. Despite this importance, and despite fierce opposition from environmentalists, plans to develop a hotel, golfing, and casino complex appear to be pushing ahead.

ISLAND TOUR
Antigua

The northern coast has a long barrier reef running – albeit discontinuously – from Diamond Reef (Salt Fish Tail) in the northwest right around to Green Island off the eastern shore. The outer edge of this barrier reef is often exposed to high swell and many shallow parts have been damaged by hurricanes, but there are still areas of healthy coral cover. Great Bird Island, which can be reached by local boat trips, is an excellent site for snorkeling, with extensive fields of branching corals in its shallow southern waters, all busy with a multitude of smaller fish, such as the stripy sergeant majors. The island is also a favored breeding ground of the beautiful red-billed tropicbird, distinguished by its red bill and two long white streamers extending back from its tail.

There are short barrier reefs in front of Nonsuch Bay and Willoughby Bay, and Green Island offers opportunities for snorkeling or diving on reefs quite close to the shore. Further round, to the south of Falmouth and English Harbours, the underwater scenery remains exciting and varied despite an absence of large coral reefs. A range of small reefs and rocky areas, some with vast boulders, offers dark crevices where nurse sharks can sometimes be seen. Stingrays often rest on sandy patches, their backs covered with sand to help them hide – even these big creatures are occasionally attacked by sharks. Lobsters, sheltering in dark holes during the day, come out at night to roam over the reef and surrounding areas looking for small invertebrates for food. They are wary creatures and usually back away from people. If they are unable to retreat into a hole they can put on quite a turn of speed, swimming backwards with powerful flicks of their tails.

Off the southwest coast is a short barrier reef known as Cades Reef. Hurricanes have taken their toll here, but there is abundant new growth and the reef hosts a great diversity of corals, including large colonies of elkhorn coral and the rare rough star coral. Considerable numbers of long-spined sea urchins inhabit the shallower waters. Watch out if you are snorkeling near these. Their slender spines pierce the skin very easily, but minute back-pointing ridges make them extremely difficult to extract. Such wounds can easily become painful and infected. Although worrisome, these urchins play a critical role on the coral reefs as algal grazers, and their abundance here may help to explain why the coral populations are relatively healthy.

Two small islands off to the west, and usually collectively referred to as Sandy Island, are surrounded by shallow reefs popular for

A cluster of long-spined sea urchins on a patch of rubble. These creatures are important algal feeders and their decline in the 1980s was disastrous for many coral reefs.

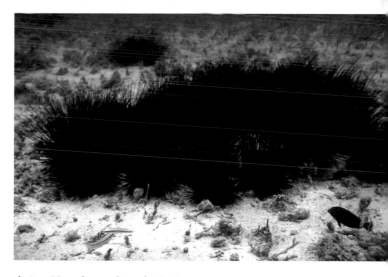

diving. Here the corals are beginning to recover from the hurricane damage of the 1990s, with extensive areas of soft corals, sea rods, and sea fans. This is also a good place for night diving, with the chance to see both sleeping fish and wide-awake invertebrates, including sea urchins, starfish, and, occasionally, the bizarre-looking basket star. The latter are related to starfish, but look more like plants with a great tangle of finely branching "arms" which they use to sieve the water and capture food from the plankton.

Some 5-10 kilometers due west of Antigua a number of shallow shoals come to within 10 meters of the surface. Rather than true reefs, these are ancient limestone structures rising out

A porkfish shelters in a bright scene of elkhorn and star coral in the shallow reefs off Barbuda.

of the surrounding sand, smothered with soft and hard corals. In places there are steep walls indented with caves, and nurse sharks often shelter in these, or at the base of the reef in about 20 meters of water. While it is possible to approach these gentle sharks, touching them is unwise – apart from causing them unnecessary stress, there is a risk they may turn round and bite. Grunts are common and in places form dense schools, often mixed with schoolmasters. Higher up in the water column ocean triggerfish are sometimes seen swimming in from the blue. Although associated with the open seas, these large blue-gray fish are known to come in to shallow reef areas to build nests in patches of sand and seagrass where they lay their eggs. They will defend these nests tenaciously if threatened.

Barbuda

This small island is in total contrast to Antigua, with only a few hotels and guest houses, and a very small local population. Reefs surround much of the island, including a nearly continuous shallow ridge formed by corals on the wave-swept east coast. There are also numerous shipwrecks along this coast, though many are largely unexplored and conditions are often too rough to allow divers access. Wide stands of healthy elkhorn coral lie to the north, as well as patches of staghorn coral. The great complexity of stony branches formed by these corals provides a home for thousands of fish. There is, of course, some fishing in this area, but large numbers of small fish, including blue tangs, black durgon, and Bermuda chub, crowd the waters all around the corals. Strong currents probably help keep the water clear and the corals healthy, but they also make snorkeling somewhat hazardous.

There are several large areas of mangrove around the island and particularly in Codrington Lagoon, which at its northern end harbors one of the largest frigate bird nesting grounds in the Caribbean. Frigate birds, sometimes called man-o'-war birds or pirates of the sea, rarely catch their own food, but attack other birds as they return to their nest, forcing them to drop or even regurgitate their food. The agile frigate birds then swoop down and catch the food in mid air.

Redonda

In addition to the two main islands, the remote and uninhabited island of Redonda lies about 50 kilometers to the west of Antigua. This steep-sided and rocky island of just some 1.5 square kilometers rises to almost 300 meters. Although it was discovered by Christopher Columbus, there was no clear claim to the island until, in 1865, a Montserratian of part Irish descent claimed it and announced his son Filipe as king. This was ignored when Antigua granted a license for mining phosphates (derived from bird droppings), which continued from the late 19th century until the First World War. The island is now inhabited only by birds (mostly brown boobies) and goats. Few visitors have dived or snorkeled in the waters around Redonda and so, while there are undoubtedly rich fish and coral communities, almost nothing is known about them.

MONTSERRAT

Temperature	26°C (Jan-Apr), 29°C (Sep)
Rainfall	1 790 mm; peak Sep-Dec
Land area	105 km² (and increasing)
Sea area	7 000 km²
N° of islands	2 (1 small)
The reefs	As a recent volcanic island, Montserrat has only a few small patch reefs, and is largely dominated by volcanic scenery. Some marine communities on the central and southern coasts have been destroyed by recent volcanic activity, but other areas have survived and indeed may be thriving.
Tourism	9 800 visitors. One dive center.
Conservation	Volcanic activity has been destructive, but has also effectively excluded humans from two thirds of the island and its surrounding waters. Parts of this *de facto* nature reserve are home to healthy marine communities, which will need more formal protection when volcanic activity comes to a close.

It is hard to imagine a worse series of events than those which have afflicted the island of Montserrat (see map, page 170), a British Overseas Territory, in recent years. In 1989 this small volcanic island was devastated by Hurricane Hugo, which destroyed 90 percent of the homes on the island. The communities and economy were just recovering when, in 1995, the Soufriere Hills Volcano began to erupt. This volcano had been dormant for the past 350 years.

Since 1995 the southern two thirds of the island, including the former capital, Plymouth, have been off-limits due to the threat posed by the volcano. Ash and debris cover most of the Soufriere Hills, and pyroclastic flows have swept down many of the valleys, even extending

Montserrat's Soufriere Hills Volcano sending out plumes of smoke and ash.

Corals and sponges can still be found in abundance off most of Montserrat's coast, and reef fish are plentiful.

into deltas offshore. (Pyroclastic flows are streams of hot rocks and ash reaching 600°C, which behave like liquids and flow down valleys at speeds of up to 200 kilometers per hour.) Gases from the volcano have created acid rain, and dust and ash have fallen across the entire island. Volcanic activity reached a new peak with an eruption in July 2003. An explosion threw ash some 15 kilometers into the atmosphere, and ash and debris fell even in the "safe" areas in the north of the island. Ash fall also affected neighboring islands, reaching as far as the Virgin Islands.

Montserrat never had extensive coral reefs – there were small patch reefs, particularly along the western coast, but elsewhere corals and other tropical marine creatures flourished on volcanic rocks. In the face of such destruction on land it has been assumed by many that the marine communities must be devastated, but, surprisingly, Montserrat is in fact still home to diverse and thriving marine life.

Of course no corals could survive the direct impacts of pyroclastic flows – where these enter the sea they cause the water to boil, and smother the sea floor with deep layers of rocks and finer sediments. Elsewhere, although there have been substantial ash falls, these have been quite limited in duration and, away from dense deposits around river mouths, the relatively heavy, inert ash has been sloughed off, leaving many corals alive and well. Even in the most damaged areas, new life quickly moves in, and corals and fish

were already colonizing larger volcanic rocks deposited along the south coast during one of the 1997 pyroclastic flows.

In the northern, "safe" area of Montserrat it is still possible to go diving, and indeed there are some very healthy marine communities with sponges and a wide range of corals, as well as patches of true reef. Fish are plentiful and larger pelagic fish are not infrequent visitors. Stingrays are sometimes seen on the sandy patches and eagle rays may fly past. Although fishing continues in the north there are relatively few fishers, and so this pressure is not too great.

It is also exciting to consider what may have been happening in the southern and central parts of the island. Here the volcano has done what few political regimes have ever succeeded in doing – it has closed off a massive area to humans. Since 1997 there has been virtually no fishing and very little other human influence. The result has undoubtedly been a burgeoning of fish and the development of thriving coral communities in many areas.

It must be hoped that, as volcanic activity comes to an end, something is made of this opportunity, with the establishment of large marine reserves. These will help to maintain fish stocks for the returning population of fishers, and may also help in the development of a thriving diving industry where divers will be able to see healthy underwater Caribbean communities and witness the colonization of new habitats left by the volcano.

A feeding stingray throws up clouds of the heavy volcanic sand as it searches for mollusks.

GUADELOUPE
ST MARTIN, ST BARTS, AND GUADELOUPE

Temperature	20-28°C (Jan-Feb), 24-31°C (Jun-Jul)
Rainfall	1 780 mm; rainfall higher May-Nov
Land area	1 735 km^2
Sea area	90 000 km^2
N° of islands	63 (11 large)
The reefs	250 km^2. The various islands have a great variety of underwater scenery, including patch reefs and coral gardens on gentle slopes, a well developed barrier reef, and steep volcanic scenery swept by strong currents that bring in larger pelagic fish.
Tourism	773 400 visitors, plus 361 700 cruise ship arrivals. The numbers vary considerably, and there are still places which are little visited. There are 16 dive centers in Guadeloupe (including one each on Iles des Saintes and Marie-Galante), plus five on St Martin and four on St Barts.
Conservation	There are protected areas on all the main islands, and these contain some vibrant reef communities, but elsewhere there are problems of pollution associated with coastal development, and overfishing.

Guadeloupe is, politically, a fully integrated Département of France – unlike the overseas territories of the UK and the Netherlands, it has full representation in national French government. And this political fact is mirrored in the reality of life on the islands, from the language to the food to the lifestyle – all French, but with a strong flavor of the Caribbean.

The political unity becomes a little confusing when looking at the geographical reality, however, as "Guadeloupe and its dependencies" includes three quite distinct island entities, and each of these has its own satellite islands. To keep the confusion to a minimum we here deal with each territory separately. The island of Guadeloupe can be seen as the political center, but the groupings of St Martin and St Barthélémy (see map, page 163) have, to a degree, a life of their own, both above and below water.

ISLAND TOUR
St Martin

The southern half of this island has already been described, being a part of the Netherlands Antilles. Although the political divide between the Netherlands and France is imperceptible (there are no border posts), there are clear cultural differences between the two halves. The French side is a little less heavily developed, but is still a very popular tourist center.

Offshore the sea floor slopes gradually downwards, but there are no dramatic walls, and only a few scattered outcrops of reef. To the north and around the island of Tintamarre are areas of more dramatic scenery with outcrops, ledges, and caves. The northeastern sites can be difficult to get to as there is often a considerable swell coming from the Atlantic, but in these areas a good range of fish life can be seen, including parrotfish, angelfish, grunts, and wrasses, with soldierfish and squirrelfish gathering in some sheltered spots. A few wrecks have been deliberately sunk as dive sites along this coast, but in fact most of the diving in St Martin takes place to the south, on the Dutch side.

St Barthélémy (St Barts)

The tiny island of St Barts has become one of the Caribbean's centers of chic, attracting mostly wealthy visitors with its quiet charm, excellent facilities, and restaurants. Towards the end of the 17th century this little island was a base for one of France's most famous pirates, Captain

Guadeloupe

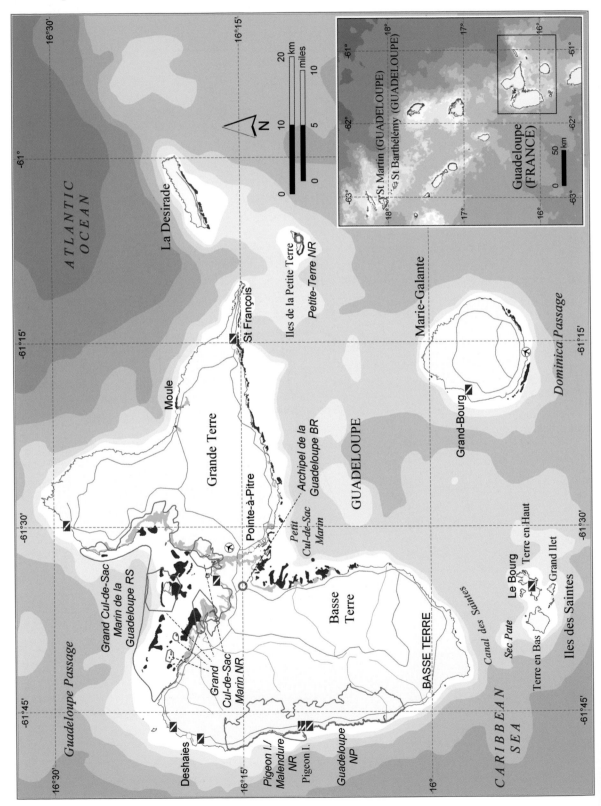

Montbars. In 1784 the island was given to Sweden in exchange for free port rights in Gothenburg, but 100 years later Sweden sold it back to France. It remained a relatively poor and undeveloped island until the 1970s.

The island lies on a wide, shallow shelf, so, as with St Martin, there are few areas below 25 meters, but the gentle slopes harbor some very attractive and healthy areas of reef. Out to the northwest dramatic scenery has formed around several tiny offshore islands, with corals growing on older volcanic rocks, including Ile Fourche and the Groupers. Here, amidst pretty scenes dominated by sea rods and sea fans, with hard corals and vase sponges, lobsters are common. Loggerhead, hawksbill, and green turtles are regular visitors. Nurse sharks can be observed on deeper ledges, and other more active species such as bull sharks may be seen.

All along the west coast corals have formed reef patches or are growing on the volcanic rock. These places have abundant fish life. Creole wrasse, blue tangs, and shimmering schools of chromis are perhaps the most abundant, but there are also considerable numbers of grunts and yellow snapper, with mahogany and gray snapper being found in some places. Spotted drums and moray eels lurk in smaller recesses. Barracuda are common, and may be seen at cleaning stations with their mouths wide open. As a signal of non-aggression to the tiny cleaner gobies, they sometimes change color, becoming much darker and showing broad bands on their sides. Tarpon are also seen in these waters, sometimes in schools of up to 100 of these massive, impressive fish.

Towards the northeast the coastline can be quite exposed, but there is some equally dramatic diving, with corals again forming around volcanic scenery, including massive submarine boulders, thickly encrusted with corals and sponges. Fire coral tends to be common in the shallower water. Further from the main island a range of shark species, including bull, blacktip, and lemon sharks, are sometimes seen.

Guadeloupe

The geography remains complex when looking at Guadeloupe – the main island is in fact two islands, while there are several other islands to the south and east. The smaller, flatter island of

The dense seagrasses of the Grand Cul-de-Sac Marin are a rich and productive nursery ground used by many reef fish. The thick foliage provides important shelter to these juvenile grunts.

Guadeloupe proper is incongruously named Grande Terre (Big Land), while the larger, mountainous island is equally strangely named Basse Terre (Low Land). These two are only separated by a very narrow channel, the Rivière Salée, lined by dense mangroves and crossed by two bridges.

Grande Terre is largely converted to sugar cane, whereas the steep mountain slopes of Basse Terre are still densely cloaked in rainforest and there are opportunities to venture on a number of trails to see the rainforest, waterfalls, and even crater lakes. The volcano of Soufrière (1 467 meters) is still active, and to avoid casualties some 72 000 people were evacuated during its last major eruption in 1976. Hot springs on the west coast are used to generate thermal electricity.

The great bay, known as the Grand Cul-de-Sac Marin, is host to one of the most important coastal wilderness areas in the Lesser Antilles – now strictly protected. Here, there are extensive mangrove forests on shore, bordering a wide lagoon with dense seagrass beds and studded with patch reefs. This leads out, in turn, to a 25 kilometer long barrier reef which drops in a steep slope to a sandy floor at between 30 and 50 meters. The interaction of all these ecosystems is considerable. Seabirds, including pelicans, tropicbirds, and noddies, roost and nest in the mangroves, while numerous fish breed amongst their roots. The seagrass beds are home to green turtles and many juvenile fish are found here. The reefs themselves house an enormous diversity of life: some 255 fish species are recorded, 157 mollusks, and 38 sponges.

Schools of Atlantic spadefish make an impressive sight on deeper reef slopes and walls. They feed on a very broad range of bottom-dwelling invertebrates.

The northern and western shores of Basse Terre are rich in life, much of it growing over volcanic rock rather than forming true coral reefs. Steep walls and piles of massive rocks are smothered in a multicolored tapestry, with sponges competing with tunicates, corals, and algae. In deeper water, and in the darker recesses, a flashlight can bring out startling colors. Under small caves the bright orange cup corals are common – they are nocturnal, but sometimes one or two polyps will have their tentacles extended during the day. The most popular place along this coast for divers is around the Pigeon Island/Malendure Nature Reserve. Also known as the Réserve Cousteau, this was a favored haunt of Jacques Cousteau. Amidst the dramatic scenery larger fish sometimes sweep in, particularly in the more exposed sites where there may be currents. Cero, barracuda, and horse-eye jacks are regular visitors, and schools of Atlantic spade-

fish are sometimes seen. Larger schools of creole wrasse are common, and, closer to the wall, damselfish such as sergeant majors, and blue and brown chromis give these reefs and walls a sense of frenetic activity.

There are well formed reefs with abundant corals along the southwest coast, and the Petit Cul-de-Sac Marin between the two islands harbors wide areas of coral reef, although these are rarely visited.

Offshore islands

To the south and east of Guadeloupe many smaller islands have their own highly distinctive characters. Rather like Grande Terre, Marie-Galante is a gently hilly island, largely converted to sugar cane production. There are coral reefs peppered around its shores, some with steeper walls and a broad range of life.

The little scattering of Iles des Saintes consists of beautiful, rugged, and dry islands with steep hills reaching over 300 meters, encompassing beautiful bays. These islands are visited by some hundreds of day-trippers from Guadeloupe, but people seldom stay overnight. Although there are few true reefs here, there are extensive areas of complex scenery with colorful rocks descending to sandy plains. Sponges, algae, and corals cover the rocks and there are many fish, including gray snapper. On the sandy plains it is possible to see the flying gurnard: these elongated brown fish are equipped with the most enormous pectoral fins. While cruising along they keep their fins folded by their sides, but when frightened open them out like a pair of enormous wings, edged with narrow lines of metallic blue. The effect is breathtaking, and of course this is what they intend – the fins greatly exaggerate the overall size of the fish, deterring predators from taking them on.

Far out in the channel between the Iles des Saintes and Basse Terre a shallow platform called Sec Pate comes up to about 15-20 meters from the surface. This isolated platform is teeming with life – often swept by currents and surrounded by deep water, it is a magnet for larger predators. Visitors are often surrounded by barracuda (or their smaller cousins, the southern sennet), cero, or jacks. The scenery is a complex of pillars and walls of coral, some undercut with hollows and tunnels.

DOMINICA

Temperature	24°C (Dec-Mar), 26°C (May-Oct)
Rainfall	1 936 mm; slightly drier Jan-Apr
Land area	732 km^2
Sea area	29 000 km^2
N° of islands	3 (1 large)
The reefs	Less than 100 km^2 of true reefs, but corals abound on volcanic rocks where there is tremendous scenery. A great place to spot frogfish, while larger fish often sweep past off the deep walls.
Tourism	66 400 visitors, plus 207 600 cruise ship arrivals. There are six dive centers along the west coast. Many tourists come for the relatively unspoiled natural environment, both above and below water.
Conservation	The Soufriere/Scott's Head Marine Reserve encompasses some of the most important marine communities and there are moves to establish other marine protected areas.

Dominica is a small mountainous island cloaked in dense rainforest and rich in wildlife. Like its neighbors, Dominica is volcanic and many of its slopes (rising to the peak of Morne Diablotin, at 1 447 meters) are too steep for agriculture.

A French colony was established here in 1632, but in 1748 the French signed an agreement with the British to leave it as neutral ground, with the local Carib population living in relative freedom. In reality there was still some settlement and there were constant tussles between Britain and France; however a Carib population was to remain, and today the island has the only Carib settlement left in the Caribbean. Britain dominated the island from the late 18th century until independence in 1978, but French influence remains strong and French creole is spoken across the island.

The high mountains attract greater rainfall than most islands, and this, coupled with the predominantly dark sand on the beaches (formed from volcanic rock), have meant that Dominica has been slow to develop tourism. This has benefited the natural fauna and flora considerably,

Hiking up into the cloud or elfin forest within the Morne Trois Pitons National Park (a World Heritage site) is an unforgettable experience: regular immersion in clouds allows life to grow on every conceivable surface.

and today the island is one of the most unchanged in the Caribbean. Locals and visitors alike are developing a growing appreciation for a vibrant natural environment set amidst spectacular scenery. Hiking trails criss-cross the island, passing through rainforests to extraordinary elfin forests and into barren volcanic

landscapes. High up, in the south of the island, lies the Boiling Lake, whose steaming surface rises and falls with the changing volcanic conditions beneath. The forests contain two parrot species found nowhere else in the world and several hummingbirds, including the blue-headed hummingbird found only here and in Martinique. There is also an impressive giant frog, know as crapaud or mountain chicken, which is considered a delicacy in the island cuisine.

Underwater, there are few true coral reefs, but all around is spectacular, steep-sided volcanic scenery, with rocks, walls, and pinnacles smothered in a rich array of corals and sponges.

Dominica

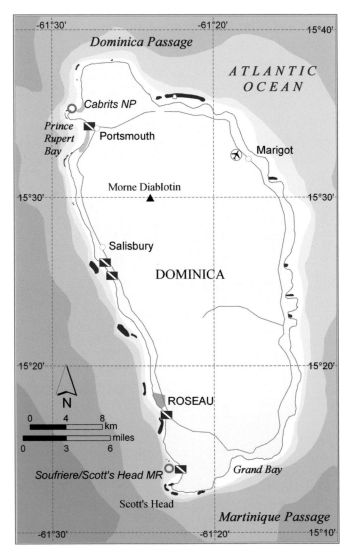

There are even underwater hot springs – with constant trickles of bubbles escaping from the sea floor between the rocks or through the sand. Up close the water and the sand feel hot and, be warned, just below the surface it gets very hot indeed. Despite the high rainfall on the island there is little sediment in the water and visibility is generally good, sometimes reaching to over 30 meters. In general, rougher conditions on the east coast preclude diving here and so little is known about corals or other marine life.

In the far north around the town of Portsmouth and Cabrits National Park the underwater scenery is varied. There are rocky areas and pinnacles, encrusted with sponges, corals, and algae. Invertebrates are common, including slipper and spiny lobsters. Off the deeper walls barracuda and cero are quite regularly seen.

Along the central section of the west coast vibrant coral gardens have developed amongst rocky scenery, with fields of sea plumes and rods broken up with steep rocks and tight crevices. Dominica is one of the best places to find some of the great masters of camouflage, including scorpionfish, seahorses, and most especially frogfish (see box). Larger fish – such as stingrays, barracuda, and snapper – are often seen, as well as an array of invertebrate life, especially at night. Brittle stars and golden crinoids, with their delicate, feathery arms, hide during the day but clamber up to high points at night to filter the water for food. Lobsters and shrimp are regulars on most reefs, their small eyes shining red in the beam of a flashlight, and in some places divers may be lucky enough to see the fabulous bush-like branches of the giant basket star.

Some of the best known underwater scenery of the island lies within the Soufriere/Scott's Head Marine Reserve. The bay here makes up one side of an older volcanic crater, and below water there are steep walls dropping to depths of hundreds of meters, high pinnacles, and some long tunnels often filled with blackbar soldier-fish. Stony corals are abundant even in shallow water, with maze coral and star corals on the steeper walls, and thickets of yellow pencil coral. Sponges, including orange elephant ear and giant barrel sponges, are abundant, while deepwater sea fans and black corals are found on the deeper, vertical surfaces. Frogfish can be spotted on the

The bright spots of the flamingo tongue snail are part of the mantle, a cloak of skin they wrap over their shells.

Moved away from its favored background, this longlure frogfish can be clearly seen. Back amongst the scenery of orange sponges it would be virtually invisible.

wall, and black margate are found in a few places. Looking out into the blue a broad range of pelagic fish make a regular appearance, including rainbow runner, cero, black jacks, barracuda, and sennet. Thankfully the importance of these waters has been realized and the marine reserve is now receiving some excellent management, with local people all highly involved in planning the future for the area.

Whale-watching is also popular just off Dominica. There is a resident population of 8-12 female sperm whales on the Caribbean side of the island, and male sperm whales come through the area, typically from December to March. These are the greatest of the toothed whales, with the females reaching a length of 12 meters, and large males as much as 18 meters. Dominica is well placed for watching sperm whales because of the very deep water just offshore. These are among the deepest divers of all the whales. When feeding they typically dive for 45 minutes, but have been known to spend two hours under-water on a single dive, and regularly go to depths of 300-600 meters (and are reported to reach 3 000 meters). In these inky depths they feed predominantly on squid, but also take fish. Closer to land, spinner dolphins are often seen in Soufriere Bay, while spotted and bottlenose dolphins also visit the area.

MASTERS OF DISGUISE

Frogfish are perhaps the best camouflaged creatures on the coral reef – almost impossible to spot without a bit of experience or some help from a more expert eye. Frogfish are typically fist sized, somewhat lumpy in shape, with tiny eyes and highly modified fins. They certainly do not look like fish, but, added to this, they are masterful color changers, taking on the hues, the textures, and even the imperfections of their surroundings. Amongst pink or orange sponges they will become pink or orange, but will be further marked with wide black circles to mimic the deep holes of the sponge. They will build into this general pattern a scattering of other markings to look like patches of algae or sand. If the same fish then moves to a different surface it will, over a few days or weeks, change color again to become green, brown, sandy, or bright red, cross-marked with fine scratches, or lumps or tufts to mimic the uneven surface of the reef.

Thus hidden, frogfish will simply sit, without moving, for days on end. They are sit-and-wait predators, and when a would-be meal comes along they have one more trick. On their heads they have a fishing line and a lure – the line is almost invisible, and the lure is like a tiny flag, which the frogfish flick forward and wiggle enticingly. Drawn in to something that looks highly edible the smaller fish come within striking distance. The frogfish then performs one of the fastest strikes of any known predator – in six thousandths of a second they shoot their mouths out and forwards, expanding them to 12 times their resting size and sucking in both the fish and a considerable volume of water. This action is too quick to see, and many times a fish is taken out from a small school without the other members of the school even realizing that it has happened.

MARTINIQUE

Temperature	21-27°C (Jan), 25-30°C (Jun-Jul)
Rainfall	1 928 mm; drier Jan-Apr
Land area	1 101 km²
Sea area	45 000 km²
N° of islands	35 (2 large)
The reefs	240 km². Reefs off the east coast are too rough for diving, but there are attractive reefs in the southwest. The northwest is dominated by coral-encrusted volcanic scenery.
Tourism	460 400 visitors, plus 201 300 cruise ship arrivals. There are 19 dive centers distributed all along the west coast and diving is a popular activity for locals as well as tourists.
Conservation	There are few active conservation measures; overfishing is a problem. There is ongoing discussion about establishing the island's first marine protected area.

The large volcanic island of Martinique, like its sister island Guadeloupe, is a fully integrated Département of France – its inhabitants are French citizens, and French law, and even some European law, applies. The island itself is quite rugged, with the steep volcanic

Schools of horse-eye jacks sometimes move in from the deeper offshore waters and will often sweep in close to divers.

slopes of Mont Pelée in the north, and more gentle fertile hills in the south offering a mixed agricultural scene, including bananas, pineapples, and sugar (from which a French "rhum" is produced). This densely populated island has many of the facilities you might associate with a European center, and also some of the problems – the city of Fort-de-France is one of the only places in the Lesser Antilles where there is a daily rush hour of commuter traffic.

Columbus first saw Martinique in 1493, although he did not actually visit the island until his fourth voyage in 1502. The Carib population remained largely undisturbed until the island was settled by the French in 1635, after which it gradually diminished through conflict and disease as well as through assimilation into the growing French population. Martinique remained a French territory for most of its colonial history.

Tourism is now a major industry in the island, and resorts make use of a great array of beaches along most of the west coast. Diving is growing in popularity, with very different opportunities in the north and the south of the country. Fishing also remains an important activity, with over 1 000 fishers on the island. Unfortunately there are few regulations, and overfishing is a major problem for the fishers as well as the ecology. Traps and nets with very narrow mesh are catching too many fish, including immature individuals, and there are few large fish left. Sadly too, there are no marine reserves where fish might find refuge. Pollution and sedimentation are localized problems and the few coral reefs in the bay of Fort-de-France have been completely degraded.

ISLAND TOUR

The eastern coast of Martinique faces a wide, gently shelving sea floor and there are coral reefs in a number of places. But these remain little known, as this coast is pounded by Atlantic swells and conditions are rarely calm enough to allow access for divers. By contrast, deep water cuts in close along much of the western shore, with a great variety of underwater terrain dominated by volcanic scenery in the north, and with more extensive coral reefs in the south.

The deep walls around the coast at the base of Mont Pelée harbor small groupers skulking in crevices and quite an abundance of parrotfish. Deepwater sea fans and black corals are widespread on the steepest walls, while sponges are plentiful everywhere. In the deeper water, particularly off the more exposed points, schools of jacks are frequent visitors, and sharks pass by.

One of the most fascinating sites in the northwest is the Bay of St Pierre itself: in 1974 a local diver called Michel Metery discovered the wreck of the three-masted *Tamaya*, one of the many ships to have been sunk during the great volcanic eruption of 1902 (see page 184). Jacques Cousteau arrived on *Calypso* soon after this discovery, and worked with Metery to find 11 more of the ships sunk in this terrible event. Many lie below 25-30 meters of water, in the realm of more experienced divers. Their molten and twisted metal bears witness to the intense heat of 100 years ago, but sea life has since moved in, smothering the surfaces with corals and sponges. Great schools of blackbar soldierfish inhabit the darker recesses.

Towards the southwest the underwater scene gives way to coral gardens. Shallower areas

Martinique

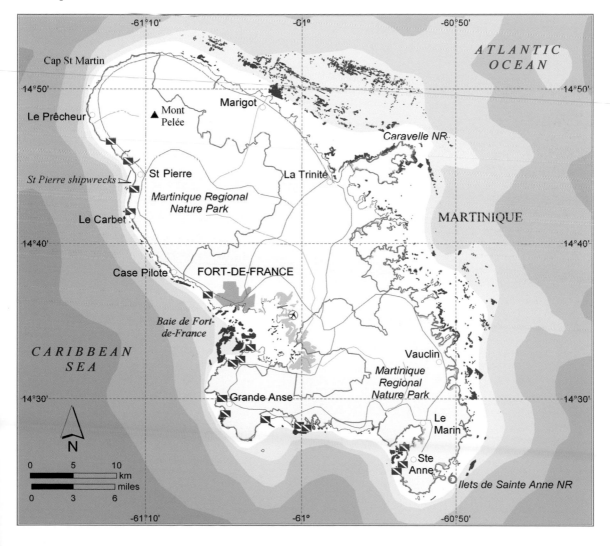

The male and female stoplight parrotfish could hardly look more different and yet all males, known as supermales (above left), have developed from adult females (above right).

A HURRICANE OF FIRE

At the start of the 20th century the town of St Pierre was a great cultural center, widely known as the Pearl of the Antilles. It was a thriving, sophisticated town with a population of 30 000, three newspapers, and a busy port, but it lay below the towering flanks of Mont Pelée, which was responsible for one of the most catastrophic volcanic eruptions in recent history.

In late April 1902, the volcano began to show its strength, with awesome firework displays spewing from its summit, and more worrying ash falls which began to cloak the city. There was considerable fear, but the city authorities were keen to reassure the population, not least because they wanted a good turnout for some important elections on May 10. At 7.50 am on May 8 the mountain exploded. A pyroclastic flow of superheated rocks and ash was hurled down the mountain slopes at speeds of perhaps 200 kilometers per hour and within two minutes the entire city was flattened and burned. Some 30 000 people were killed, but there was one survivor – a prisoner in a windowless cell.

Just offshore there were 18 boats in the bay: 17 were destroyed and sank, but the British steamship *Roddam* stayed afloat and a few people survived, suffering tremendous burns. They offered the only eyewitness accounts. Assistant Purser Thompson wrote that "the mountain was blown to pieces. There was no warning. The side of the volcano was ripped out and there was hurled towards us a solid wall of flame. It sounded like a thousand cannons. The wave of fire was on us and over us like a flash of lightning. It was like a hurricane of fire... The town vanished before our eyes."

Today there is a small village at St Pierre, but many of the ruins from 1902 remain as they were, and 12 wrecks from the explosion remain offshore.

are dominated by sea rods and sea fans, but there are also steep walls decked with brain and sheet corals, as well as whip corals and some black corals. Parrotfish are quite common, particularly stoplight and princess parrotfish, making this a great place to get to grips with the extraordinary differences in color between juveniles, adults, and the dazzling "supermales". Like their relatives the wrasses, parrotfish change sex as they grow older, beginning life as females and later becoming males – so in its lifetime a parrotfish will move through all these color phases. Larger groupers are rare in these waters, but the occasional coney or red hind may be seen.

Bicolor damsels, as always, are abundant. These striking little fish live in small groups with a complex social hierarchy – watch them for a while and see an almost continuous succession of chases and challenges between individuals. Like many damselfish they breed throughout the year save during the coolest months. The males try to tempt females into a nesting area, usually a small crevice, or even a discarded conch shell. The eggs are stuck to the nest floor and are then stoutly defended by the male for the three or four days it takes for them to hatch. Many fish, such as bluehead wrasse, love to eat damselfish eggs, but these small lone males will fight hard to chase off any intruders, so it is only when a concerted approach by a school of egg eaters takes place that they are defeated.

St Lucia

Temperature	24-27°C (Jan-Feb), 26-30°C (Aug-Oct)
Rainfall	1 500 mm; wetter Jun-Nov
Land area	605 km²
Sea area	15 000 km²
N° of islands	9 (1 large)
The reefs	160 km². Healthy coral scenery abounds in St Lucia, although it is generally only the west coast which is accessible. Living amongst gentle coral gardens and dramatic walls is a great diversity of fish life.
Tourism	250 100 visitors, plus 489 900 cruise ship arrivals. There are ten dive centers along the west coast, with most diving activity focused around Anse Cochon and Soufriere.
Conservation	Early efforts to establish marine reserves in St Lucia were largely ignored by local fishers, but today the Soufriere Marine Management Area is leading the world in integrating conservation with fishing and tourism. Underwater the results are spectacular.

The lush green island of St Lucia is regarded as one of the most beautiful in the Lesser Antilles, with its forested hills, sheltered natural harbors, and isolated beaches. Despite its serene beauty today, the island has a long and checkered history. The original Carib peoples fought hard against the first European arrivals who attempted to settle in 1605 and 1638. In 1650, however, France established a settlement, and the Carib people were quickly decimated. From this time until 1814 Britain and France struggled almost continuously over ownership of St Lucia, which changed hands more than any other island in the Caribbean – French one decade, British the next. Today, although English is the official language, French creole is still widely spoken. Tourism is the mainstay of the economy, although bananas provide important export earnings, and there is a small industrial base in the south.

Volcanoes have built this landscape, and the south of the island is famous for its scenery. Close to the small town of Soufriere there is a wide caldera known as the Qualibou Depression (Qualibou is a Carib word meaning "place of death") – this is the area of Sulphur Springs, producing superheated water and steam, pools of boiling mud, and hot sulfurous vapors. Despite this dramatic scenery and sulfurous air, the last significant activity was a minor eruption in 1766 which threw out ash over a wide area. In the Diamond Gardens – in the grounds of the botanic gardens first built in 1784 for the French King Louis XVI – there is a hot water spa with a waterfall and pond.

Perhaps the best known volcanic features on St Lucia are two enormous lava domes which lie right on the coast – the Pitons. Gros Piton (reaching to 760 meters) and Petit Piton were actually formed before the large Qualibou caldera; cloaked in green, they rise precipitously

With its array of lumps, tufts, and tassels the spotted scorpionfish has a formidable camouflage. A sit-and-wait predator, it relies on being almost invisible, waiting to engulf small fish in its voluminous mouth when they stray too close.

St Lucia

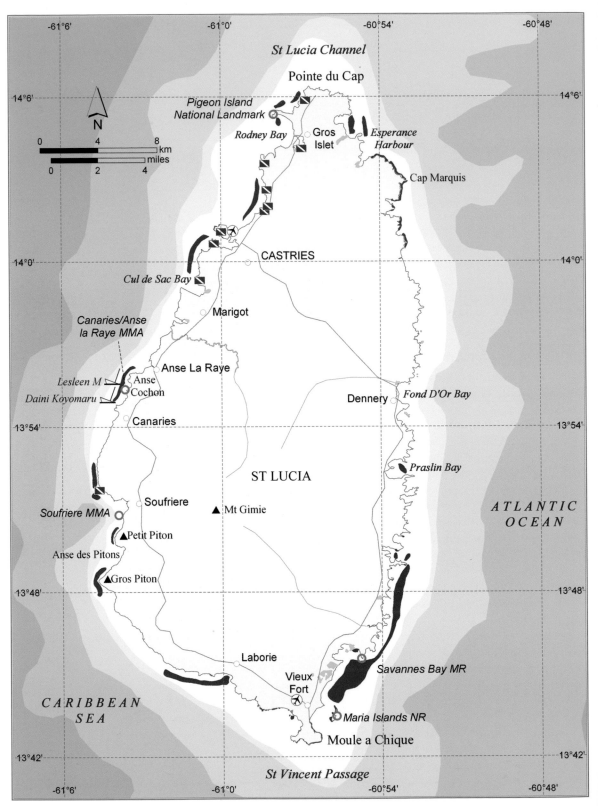

from the sea in two great towers. Below them the underwater scenery continues in dizzying walls, with spectacular opportunities for exploration.

ISLAND TOUR

Access to the sea is largely restricted to the western shores, where there are some coral reefs, but where the most extensive marine life is found growing directly on volcanic bedrock. Coral gardens begin in quite shallow water around Anse Cochon; however the most prolific life is found on the steeper slopes and walls. The colors are remarkable, with a great diversity of encrusting, rope, and barrel sponges, but also a host of both soft and hard corals. Amongst this scenery juvenile spotted drums are quite often seen, gently twisting and turning below the corals with their high dorsal fins waving like flags. Goatfish are fairly common amidst the corals, particularly near sandy areas. Up close it is possible to see the strange prehensile barbels on their chins (they are supposed to look a little like goats' beards, hence their name). These are packed full of taste receptors and the goatfish use them to rifle through the sand in search of worms, mollusks, and crustaceans.

Harder to spot is the occasional scorpionfish. These predators are near invisible, with their mottled brown or reddish skin covered in fine tufts and growths. They lie in wait on the sea floor until a smaller fish passes close enough to be engulfed in their enormous mouths. They are also well defended against any fish which might try to eat them – the spines of their dorsal fins hold a powerful poison.

Two large wrecks have been sunk for divers in the area of Anse Cochon – the 80 meter *Daini Koyomaru* was sunk in 1996 and is already populated with quite a range of fish. Life is even more abundant on the *Lesleen M*, a 50 meter freighter sunk in quite shallow waters in 1986. Lying in a sandy arena populated by garden eels, this wreck is now thickly encrusted with sponges and corals. Branching hydroids – which are closely related to fire corals and build simple feather-shaped colonies – are common on the surfaces of wrecks, and on some it is quite possible to see the individual polyps. Some

The Pitons are volcanic plugs which formed in the center of lava domes that developed on St Lucia 200 000-300 000 years ago. These spectacular formations continue underwater in vertical walls, smothered in life.

species have stinging cells of sufficient strength to produce a painful sting even in humans. As is so often the case, squirrelfish and blackbar soldierfish have moved in and populated parts of these wrecks.

The Soufriere Marine Management Area (see page 188) offers some of the best opportunities to see thriving coral reefs in all of the Lesser Antilles. Although quite heavily dived, and also fished in parts, there is an abundance of fish, while the reefs themselves rival any others in the Caribbean for color and vitality. Soft coral gardens with occasional pillar corals typically descend quite quickly into a steep slope or reef wall where plate corals and a profusion of

Branching hydroids, like corals, are colonial animals with separate polyps. They are in the same class as the fire corals and, like them, can produce a painful rash on contact with bare skin.

sponges abound. These include the sculpted forms of azure vase sponges, yellow tube sponges, red rope sponges, and some very large barrel sponges. Deepwater sea fans and sea whips inhabit the deeper slopes.

Adding a sense of frenzied activity to this scenery, fish are superabundant. There are often large schools of creole wrasse, and bar jacks are common. In amongst the corals damselfish, bluehead wrasse, Spanish hogfish, and butterflyfish are to be seen everywhere. Larger fish such as mutton snapper, black margate, and Spanish grunt are found in the marine reserves and there is a resident school of crevalle jack. These are large deep-bodied jacks with especially powerful jaws – they are inquisitive and may form a tight swirl around a diver or snorkeler for a good look.

On the more exposed sites around Soufriere there can be strong currents, often allowing more experienced divers to be swept past the fabulous scenery on drift dives. The incredible towers of the Pitons make an extraordinary backdrop to this area, but their contours actually continue underwater in a few locations, and vertical walls and breathtaking pinnacles rise sheer out of deep blue waters to within just a few meters of the surface. Seahorses and frogfish are sometimes spotted by the more observant divers when swimming along these walls.

Rough conditions and currents generally restrict access to waters off the eastern side – where true coral reefs are known to exist but for the most part are poorly known. This coast is undeveloped and very beautiful, with sheltered bays and quite a number of small islands just offshore, many of which are home to seabird nesting colonies. Perhaps the most interesting of these islands are the Maria Islands, a nature reserve and home to the St Lucia whiptail, a ground lizard found nowhere else. This creature is famous not only for its rarity, but because it bears the colors of the St Lucia flag, dark on top, with a bright yellow belly and a striking blue or turquoise tail. In the sheltered waters behind the Maria Islands extensive areas of coral in shallow water make for excellent snorkeling.

SOUFRIERE MARINE MANAGEMENT AREA

Over a wide area around the small coastal village of Soufriere and the dramatic scenery of the Pitons, a marine protected area is setting world standards in the management of marine resources in the face of the apparently conflicting needs of fishers, tourist operators, divers, and the environment. Since 1992 a tiny marine reserve off Anse Chastenet has been roped off with buoys by an adjacent hotel and fishing has been actively excluded: by 1995 there was a notable increase in the numbers of parrotfish and snappers within this small reserve, and individuals were reaching much larger sizes.

While this was going on there were also much wider efforts to develop better protection for this coastline, involving a long period of consultation with local people. It was critical to get support from the fishing community that relies on the coastal resources – without such cooperation many marine reserves have totally failed. Other needs were also noted – divers, of course, like pristine reefs with lots of large fish, while the area is popular for yachts, so it was important to keep areas where yachts were allowed to moor.

The critical point was the realization that, with a reasonable approach, everyone would benefit. A plan was drawn up with four marine reserve areas where no fishing of any sort was permitted, alongside yacht mooring areas, multiple-use areas, and also fishing priority areas. Some fishermen objected to losing more than a third of their former fishing grounds, and were provided with compensation for a short period. But over just a few years the changes have been truly remarkable. There are now four times more fish in the marine reserve areas than there were in 1995, and these are spilling over into the surrounding areas, which have three times more fish than before. The fishers are finding that their catches have gone up enormously – they are catching more fish for less effort, despite the loss of some fishing areas. Tourists, of course, are finding some really excellent diving, while the fees charged to divers, snorkelers, and yachts are paying for the management of the whole scheme.

Today, most of the community is supportive of the entire scheme and their ideas are being copied in the newly established Canaries/Anse la Raye Marine Management Area immediately to the north, as well as in new reserves in other parts of the Caribbean.

ST VINCENT AND THE GRENADINES

Temperature	24°C (Feb-Mar), 26°C (Jun-Nov)
Rainfall	1 928 mm; wetter Jul-Dec
Land area	390 km²
Sea area	38 000 km²
N° of islands	65 (8 large)
The reefs	140 km². St Vincent itself has spectacular volcanic scenery, though few true reefs, and has fabulous black coral communities. Coral reefs are widespread throughout the Grenadines.
Tourism	70 700 visitors, plus 76 500 cruise ship arrivals. There are 12 dive centers on five islands. Most of the tourism is focused on the Grenadines, and a little in southern St Vincent. Diving is popular in all tourist areas.
Conservation	St Vincent's underwater life has been protected more by a lack of human pressures than by any positive attempts at conservation.

St Vincent is a smaller island than its neighbors to the north, but has the same grandeur, with its own high volcano and forested hills. Quite unlike its neighbors, the country also contains its own archipelago of small islands to the south. These are the St Vincent Grenadines, and make up about two thirds of the entire Grenadines island group (with the remainder belonging to Grenada)

St Vincent's Carib population was quite successful at deterring settlers – some slaves were shipwrecked here in 1673, but the first concerted European efforts at settlement did not come until the mid 18th century. Because the island remained relatively isolated and "free" it became a focus for both Caribs and slaves fleeing from some of the other islands. There was a considerable ethnic mixing, creating a group known as black Caribs, some of whom still remain in the northern villages.

There are many hiking trails on St Vincent – the denser areas of rainforest are home to the rare St Vincent parrot, found nowhere else in the world – but perhaps the most demanding are up the Soufriere Volcano (1 234 meters). Paths wind up through thick rainforest and into the clouds and a quite desolate landscape. Soufriere erupted just the day before Mont Pelée in Martinique, in 1902, and some 1 600 people were killed by pyroclastic flows. Predictions of another eruption in 1979 led to the evacuation of 17 000 people. This time a massive cloud of ash, with a diameter of 140 kilometers, was thrown some 18 kilometers vertically into the atmosphere. Once again massive pyroclastic flows of superheated rock and ash raced down the mountain slopes and reached the sea in the northwest.

The Grenadines to the south are difficult to access for most people, but they have achieved renown among yacht-based travelers as well as those seeking a particular type of Caribbean experience. These are small isolated islands, with scrub-covered hills, lined with untrodden white sand beaches All around there are coral reefs.

The dark shades of feather black corals are easily overlooked, but form majestic plumes on steep walls and deeper slopes of the reef.

St Vincent and the Grenadines

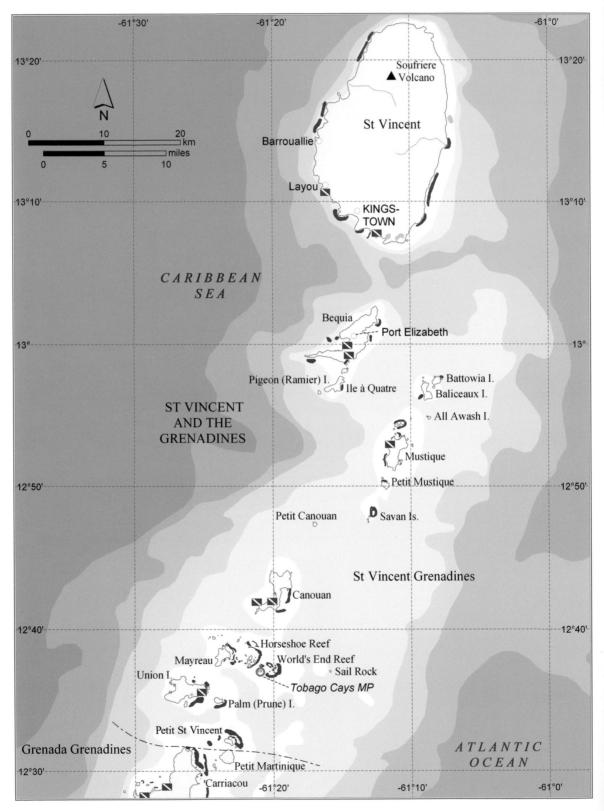

ISLAND TOUR
St Vincent

St Vincent has some of the least impacted marine communities in the Lesser Antilles – there are a few areas of true coral reefs, but the most spectacular scenes are dark volcanic seascapes, with steep walls along much of the western coast smothered in marine life, often dominated by sponges. There is little pollution, and with few divers and fishers, there is little damage from boat anchors.

In many places shallow shelves above the steep walls are home to gently sloping coral gardens dominated by soft corals, but with very large brain corals, and patches of finger corals. One of the smallest and shyest of the Caribbean angelfish, the cherubfish, is found here. It reaches about 6 centimeters, but up close is a very beautiful rich blue suffused with violet, and with a golden face and chin. The sandy areas above the walls are famous for one of the most bizarre Caribbean fish – the batfish. Somewhat flattened and almost circular from above, they have stiff pectoral fins projecting out from flange-like extensions to their bodies, while their pelvic fins are tucked up underneath. These two pairs of fins are used just like legs to walk on the bottom. Batfish are related to frogfish, and like them they are ambush predators. They tend not to move even when divers or snorkelers approach, assuming that their camouflage will make them invisible.

The steep walls are home to other oddities, including seahorses and frogfish, as well as the usual shimmering hordes of chromis and damsels, and the occasional brilliant purple and yellow flash of the fairy basslet. In amongst the thickly decorated walls, divers are likely to see a range of lobsters, including spiny and slipper lobsters, and also the much rarer red-banded lobster, whose long antennae are banded in red and golden orange, and whose bodies are similarly mottled.

Black corals are abundant: unlike most stony and soft corals, they have no algae living within their tissues. They are not reliant on sunlight and thrive in deeper water and on steep, dark walls. Black corals are so-called because they build a stiff skeleton from a dense black protein. They come in a great array of forms, from bushes, to fans, to feathery plumes, and their body tissues contain pigments of green, red, brown, gray, or orange so, although the skeleton lends a sombre tone, they never appear truly black. When dead, however, the skeletons retain a deep black and can be crafted into valuable jewelry. This has led to the over-harvesting and virtual disappearance of black corals in many countries.

St Vincent Grenadines

Unlike the main island of St Vincent, the Grenadines are scattered across a relatively shallow shelf and coral reefs are widespread, growing on an undulating sea floor. In many places there are submerged boulders or larger rock outcrops which rise up from sandy floors offering steeper walls, and small caves and overhangs. Wide areas, especially in shallower water, are dominated by the wafting shapes of sea plumes and sea rods, and there are extensive fields of common sea fans. Butterflyfish and the occasional vibrant hues of a Spanish hogfish are often seen, along with the stealthier figures of trumpetfish. There are also densely packed schools of bluestriped and French grunts. Stony corals such as boulder star corals and sheet corals, as well as tunicates, join the soft corals on the steeper slopes. Large French and queen angelfish are often seen moving through the corals, and queen triggerfish are regular visitors. Southern stingrays are found on the adjacent

Boat-building is an important local industry in the Grenadines, and these small colorful craft are widely used by fishers throughout the islands.

Creole wrasse often swim over the reef in large schools: here part of a group has broken away from the main school, with a larger and more colorful male above a group of females.

sandy areas, and turtles – especially green and hawksbill turtles – are widespread throughout the Grenadines.

Currents are common in St Vincent and particularly around the smaller islands of the Grenadines. In many places visitors are taken on drift dives. The sea around rocky outcrops is often seething with fish, encouraged by the shifting currents and the availability of plankton. Brown and blue chromis swim high above the corals, sometimes joined by sergeant majors, in a scattered frenzy of activity, searching the water around for particles of food. But if a larger fish like a barracuda appears they turn, as one, and dive head down towards the sea floor. Larger planktivores such as creole wrasse often join the activity, and occasionally there may be schools of the streamlined silvery blue forms of boga.

Nurse sharks are often found resting on the bottom in the more remote sites, and more active sharks, including Caribbean reef sharks and blacktips, are also regular visitors. These two species are closely related, and look very similar, although the blacktip has quite clear black edges to its fins and upper tail. Blacktips tend to be more active sharks, often seen near the surface, where they are voracious fish-eaters, and even take other, smaller sharks, although they do not pose a threat to people.

Bequia is one of the largest of the Grenadines and, before tourism took hold, the major activities of the island included fishing and shipbuilding. A "traditional" whaling industry still operates from the island, working from small wooden boats using only hand-held harpoons. There is, however, considerable controversy over this whaling. The whalers first target humpback whale calves and, once one is caught, pursue the mother who stays nearby in an attempt to protect her offspring. The targeting of calves is actually illegal under international law, but this has never been properly challenged and so the practice continues. Typically, only two whales are caught per year, but in 2002 the permitted quota was increased to four. We know so little about these whales that it is not possible to state whether this catch is sustainable. We do know, however, that this activity has sparked considerable protest and has probably deterred many visitors from coming to the island.

To the east of Mayreau lies a scattering of uninhabited islands known as the Tobago Cays, each with a tiny, bright beach, and together providing one of the most picturesque anchorages in the southern Caribbean. All around are coral reefs, including Horseshoe Reef and World's End Reef, which are the largest reef systems in the country, and provide the cays with shelter from the prevailing wind and waves. The water around the reefs is clean and clear, and they maintain some beautiful scenery, where mahogany snapper are common and surgeonfish abound. Unfortunately, anchor damage and nutrient pollution arising from the many hundreds of yachts which pass through each year, together with overfishing, has spoiled the corals in some areas and reduced the numbers of larger fish. This area is now part of the Tobago Cays Marine Park, and there are growing efforts to strengthen conservation, in collaboration with the visitors and the local fishers. If these efforts pay off then it might be hoped that things will begin to turn around for these reefs.

Two small reefs, just to the west of the tiny island of Petit St Vincent, each marked with a tiny sand cay too small to support plant life, are both very healthy, with fields of coral descending to depths of over 30 meters. There are still a few beautiful stands of elkhorn coral in the shallow areas and a host of different parrotfish – stoplight, redband, princess, queen, and striped – pick their way around the reef, occasionally diving down to scrape algae from the rocks.

GRENADA

Temperature	25°C (Jan-Feb), 26°C (Mar-Dec)
Rainfall	1 909 mm; wetter Jun-Nov
Land area	367 km^2
Sea area	25 000 km^2
N° of islands	58 (4 large)
The reefs	150 km^2. There are wide areas of reef around Carriacou in the north, and to the southwest of Grenada. Around Isle de Ronde the scenery is dominated by volcanic rock, but still teeming with life.
Tourism	123 400 visitors, plus 147 400 cruise ship arrivals. Seven dive centers on Grenada and two on Carriacou, with diving widespread through the Grenadines, and focused in the southwest of Grenada.
Conservation	There have been plans for some time to develop a comprehensive system of protection for Grenada's reefs, but to date very little has been achieved.

The southernmost island of the Lesser Antilles, Grenada is another hilly, richly fertile, volcanic island. Also known as the Spice Island, Grenada is a tapestry of small farms growing a wealth of crops, including nutmeg and mace, but also limes, coconut, mangos, passion-fruit, guava, tamarind, and of course the ubiquitous banana. The steeper and more remote parts are still heavily forested, especially the volcanic peaks which run down the center of the island. To the north of Grenada itself, the country is also host to a scattering of small, hilly islands, the Grenada Grenadines.

The northernmost peak on Grenada, Mt St Catherine (840 meters), is a dormant volcano and there have been no eruptions in the last 1 000 years. However, Grenada is host to the Caribbean's only active submarine volcano. Kick-'em-Jenny lies in deep water about 8 kilometers north of Grenada. She was unknown until 1939 when numerous people in northern Grenada watched a great cloud rising 250 meters above the sea. There have been a number of eruptions since, as recently as 2001, but only the eruptions of 1939 and 1974 actually broke the surface. The depth and shape of the volcanic crater is constantly changing – it now lies at about 183 meters below the surface, having been at about 160 meters in the 1980s and 235 meters in 1962. It seems possible that Grenada may be host to a new island in the coming years.

Columbus saw Grenada on his third voyage in 1498, but European settlement took many years. The local Caribs deterred the first (British) attempts on the island, but in 1650 a French company established a small settlement. From the late 18th century until independence in 1974 Grenada was under British rule. Although there were struggles in the early years following independence, including a bloodless coup, a brief period of neo-communism, and a US invasion, Grenada is now stable and tourism has become the major industry. Most is focused in the south-

A dense school of mutton snapper and bluestriped grunt waiting out the day before heading out across the reef to feed at dusk.

west where the capital, St George's, provided an excellent harbor for the early Europeans, and today remains a remarkably picturesque town with small, brightly painted, red-roofed houses ascending the hillside.

Underwater, Grenada enjoys a mix of coral reefs and volcanic slopes housing coral communities, with large numbers of fish and some very healthy coral scenery. Around the islands of the Grenadines there are more extensive coral reefs in many places. Although there are increasing efforts to protect the natural environment, and a large number of national parks and nature reserves have been proposed, more needs to be done to get these up and running.

ISLAND TOUR
Grenada

Few divers or snorkelers get to visit the east coast, which is swept by wind and waves, but there are some rich marine communities, particularly in the more sheltered bays all along this coastline. There is also some very pretty reef scenery around the northeast tip of the island, with wide stands of staghorn and elkhorn coral. These areas are included in the Levera National Park.

To the north of St George's lies an area sometimes referred to as the Grenada Marine

Grenada

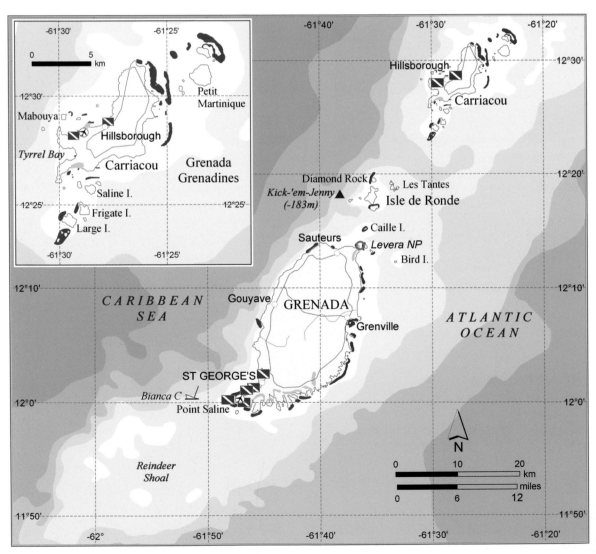

Park, or Molinere Reef, where efforts have begun to protect some very beautiful and important marine communities. Here, the shallower reefs tend to be dominated by the gently wafting forms of soft corals, but rising above these shifting scenes are the solid towers of pillar corals. Small fish are abundant, including damselfish and the bright metallic flashes of the blue chromis. Schools of creole wrasse are very common in this area, often numbering several hundred fish which swarm in over the reef and seem virtually to ignore divers, while spotted moray eels shelter or patrol amongst the corals. Beyond the shallow coral gardens there are areas of steep slopes and walls where deepwater sea fans are common, and the occasional larger grouper can be seen.

A long offshore reef system runs southwest from St George's towards Point Saline, punctuated by canyons and sand patches with countless fish. In the east this reef is dominated by finger, boulder, and pillar corals, with some brain and soft corals. Towards the west the sea plumes and sea rods begin to dominate. Shy and barred hamlets pick their way amongst this changing scenery, and the occasional gangs of bright yellow female bluehead wrasse swim over the corals. Lobsters are common in the darker crevices.

Further offshore, other reefs are often visited by hawksbill turtles and larger predators, including barracuda and flashing silvery schools of southern sennet. Green morays explore the reef slopes and small groups of Atlantic spadefish sometimes come up to investigate divers.

The southern shore of Grenada is swept by strong currents. Here there are a number of reefs as well as broad sandy areas, presenting a generally undulating terrain with gardens of sponges and soft corals. The cushion sea star – the Caribbean's largest starfish reaching to 40 centimeters in diameter – is found on these sandy areas. They feed on a variety of foods, but especially bivalve mollusks – like all starfish they are capable of extruding their stomachs and can digest their food outside their bodies. Southern stingrays are also quite often seen in the sandy areas, and eagle rays occasionally fly over. On the reef areas tarpon are found, with their huge metallic scales looking like armor-plating. Black durgon are abundant, making the most of the water movements to feed on the plankton. On

EIGHT LEGS, EIGHT NOSES, AND 20-20 VISION

The octopus is one of the more extraordinary denizens of the reef. Octopuses are very shy creatures with excellent eyesight, which means that they are difficult to spot. Even so they are quite commonly seen on the volcanic slopes of the southern Lesser Antilles. As they are approached they tend to disappear very quickly into the smallest holes, so on seeing one the best reaction for a hopeful octopus watcher is to keep absolutely still.

Octopuses are related to mollusks such as conch and oysters, but millions of years of evolution have led them through some quite extraordinary changes to make them what they are today, the most intelligent and quick-witted of all invertebrates. The word octopus means "eight legged", and octopuses belong to a class of mollusks known as cephalopods (meaning "head foot"), along with the squid, cuttlefish, and nautilus. Connected directly to their feet (or legs) is a large head, bearing a pair of eyes which, anatomically speaking, are remarkably similar to a mammal's and give octopuses excellent vision.

On coral reefs, octopuses are bottom dwellers and they use their legs, which are equipped with powerful suckers, to walk across the reef. These legs are also tipped with powerful "chemoreceptors", effectively providing a sense of smell, or taste, that can be poked into the holes of the reef in search of food, which is typically crustaceans and mollusks. Between their legs is a mouth with a powerful beak that easily cracks open even quite thick shells and skeletons.

These wonderful, inquisitive creatures are also masters of disguise. They regularly change the texture of their skins, while they also have specialized cells known as chromatophores which can be expanded or contracted to create different colors. The Caribbean reef octopus is regularly seen changing through various shades of blue, pink, green, brown, or red, but is more typically an ingenious mottling of colors perfectly adapted to its surroundings. Octopuses also use their color-changing skills to communicate, flashing from white to almost black if threatened, or sending pulses of color across their bodies like a display of lights when communicating with one another.

Common octopus.

A bipinnate sea plume and a common sea fan provide a stage for the gentle movements of a juvenile spotted drum.

the offshore reefs there are also nurse sharks – gentle giants which can reach 4 meters in length, and are regularly seen at 3 meters. Caribbean reef sharks also come into these reefs.

There are several wrecks close to St George's, including the 200 meter long former cruise liner the *Bianca C*, which sank here in 1961. Most of these wrecks have been in place for several decades and are heavily encrusted with corals and sponges. The pretty, delicate forms of white telestos are abundant on several of the deeper wrecks – these small corals are actually considered a fouling organism on boats, and rarely grow on reefs. On shipwrecks, however, they make a beautiful decoration, with their short fronds smothered with frosted white polyps. Larger angelfish often take up residence in wrecks, along with the occasional groupers or large margates.

Grenada Grenadines

Carriacou is the largest of the Grenadines, a dry hilly island with long, white sand beaches. The local population includes a number of highly skilled boat builders who use traditional techniques (with no plans and no power tools) to build highly distinctive fishing boats. There are coral reefs all around the island, including a barrier reef along the east coast, and around most of its offshore satellite islands. Diving is

popular, mostly taking place off the more sheltered western shores, and there are also good opportunities for snorkeling in the shallow waters.

The reefs are quite varied. Around Frigate Island there is a great diversity of hard corals, including symmetrical brain, boulder star, great star, sheet, smooth flower, and cactus coral. In general, however, soft corals dominate a large number of the reefs. Identification of many of these corals can be tricky, but it is possible to distinguish the more feathery branches of the various species of sea plumes – some of these can grow into vast bushes reaching 2 meters in height. Sea rods tend to have upwards pointing, less finely branching stalks, and closer inspection of the retracted polyps shows that some have smooth surfaces covered in simple round pores (porous sea rods), while others have narrow slits (slit pore sea rods), and others have raised polyp bases (warty sea rods). Sea whips tend to grow a little like sea rods, but most have their polyps arranged along the angular edges of the coral stalks. Amongst the regular crowds of fish in these coral gardens it is quite common to see smooth trunkfish – these, the commonest of the boxfishes, are relatively slow-moving fish which have developed a strong armor to protect themselves from predators. Apart from their fins, tails, and eyes, their bodies are tightly armored and inflexible. They tend to move slowly over the bottom in search of small invertebrates.

Apart from Carriacou and the nearby Petit Martinique, the Grenada Grenadines are uninhabited. Much closer to Grenada, the Isle de Ronde and its neighbors are a series of dramatic islands of obvious volcanic origin. Reefs have had little time to form around some of these, and indeed geologists believe that Caille Island may have formed as recently as 1 000 years ago. The submarine scenery includes steep walls, but also shallower shelves broken up with vast volcanic boulders. The waters are very clear here, and strong ocean currents often sweep the islands. Barracuda and schools of horse-eye and bar jacks are common, and nurse sharks are regular visitors. Midnight parrotfish are also found here – reaching to 80 centimeters, these are one of the parrotfish which keep the same colors throughout their lives, a rich dark prussian blue with patches of vivid azure around their heads.

BARBADOS

Temperature	23-27°C (Jan-Feb), 26-29°C (Jun)
Rainfall	1 180 mm; wetter Jul-Nov
Land area	440 km²
Sea area	186 000 km²
N° of islands	1
The reefs	Less than 100 km². Although fringing reefs on the west coast are largely degraded, the south and west coasts have a bank barrier reef system lying further offshore which is covered in healthy coral communities and numerous fish. The east coast reefs are largely inaccessible, but include some important fish life.
Tourism	507 100 visitors, plus 527 600 cruise ship arrivals. There are 13 dive centers operating all along the west and south coasts. Interest in diving is growing among local people and visitors. A large number of ships have been sunk as wreck diving sites.
Conservation	Although there are two named marine parks these have been relatively ineffective to date, although it is to be hoped that growing interest in the marine environment may soon change this.

First discovered by the Portuguese, the name Barbados roughly translates as "bearded ones" and is probably derived from the large numbers of weeping fig trees, with strange, shaggy, aerial roots hanging down from their branches. The island was uninhabited when English settlers first arrived in 1627, but there had been an Amerindian population some time before this – perhaps wiped out by slave traders or European diseases. Unlike the other islands, there was never any fighting over Barbados – it was claimed for Britain and remained British until independence in 1966.

Barbados sits quite apart from the other islands of the Lesser Antilles. It is gently hilly, formed from sedimentary rocks rather than the raw mountains produced by volcanoes. Most of the surface rock is ancient coral reef reaching to over 90 meters thick, and in a few places deep caves have been eroded into this limestone.

The east coast of Barbados is still quite undeveloped; pounded by surf and swept by currents, it seems a world apart from the tourism-dominated west coast.

Once the island was covered in thick forests, but these are now all gone and the island is dominated by agriculture, mostly sugar cane. Tourism, however, is the island's greatest success and has led to the considerable development of the western and southern coasts, where very large numbers of visitors come to enjoy the climate, the beaches, and the golf. In stark contrast, the east coast, with steep hills and wild, wave-swept beaches, remains secluded, quiet, and very beautiful.

There is a growing interest in the diving opportunities around the island. Although a small marine park has existed on the west coast for many years, and the waters of Carlisle Bay

Barbados

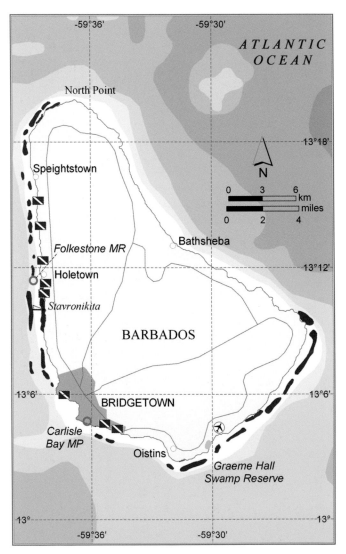

are also referred to as a marine park, neither has offered effective protection and these areas are still being damaged by fishing and boat anchoring. Things appear to be changing, however. Mooring buoys are being established on some dive sites, and the growing interest among the people of Barbados in their marine heritage may mean that things are going to improve in the coming years.

THE CORAL REEFS

Along Barbados' south and west coasts a series of bank barrier reefs rises up from a sandy floor at about 40 meters and comes to within 15 meters of the sea surface. Typically around a kilometer offshore, and well below the surface, these reefs have remained very healthy and are smothered in a wealth of life, with very large numbers of stony corals. Boulder star, great star, finger, and a range of brain corals abound. There are a few patches of elkhorn in the shallower areas, and pillar corals rise up in great towers. The steeper slopes and walls tend to be dominated by sheet corals, and there are also black coral bushes and the orange flashes of elephant ear sponges.

Fish are abundant on all the reefs. The more heavily visited sites have large numbers of smaller fish – chromis and bicolor damsels are abundant, while French grunts, Spanish hogfish, and the occasional parrotfish brighten the scene. Larger angelfish are also often found on these reefs, and hawksbill turtles are common visitors – both angelfish and hawksbill turtles regularly feed on the plentiful sponges. Schools of Bermuda chub are found on many reefs and have become used to being fed in some places.

Trumpetfish are abundant, particularly where soft corals are dominant – these slender fish are active predators, often using the cover of sea rods to ambush their prey. They are also masters at shadowing – swimming close beside another fish such as a parrotfish to avoid being seen, then lunging forwards when their quarry comes within reach. Unlike other predators such as groupers, they are not attractive to fishermen, and some scientists have suggested that they may be more abundant on certain reefs precisely because other predators have been overfished, leaving the trumpetfish to fill the niche.

The more remote reefs of the northwest and south coasts tend to be frequented by larger fish, including occasional schools of barracuda, and sometimes cero or dog snapper. Large deepwater sea fans and giant barrel sponges are also common.

Most of the nearshore fringing reefs in Barbados have suffered badly from pollution, so there are few opportunities for snorkeling. Off the south coast shallower reefs come a little closer to the shore, but these are generally still below the range of snorkeling, and there can be currents in these areas.

The east coast can only be accessed under very calm conditions, which are rare. In general the topography is low lying and smooth due to the constant pounding by Atlantic swells, but these areas have some corals, and are home to a considerable diversity of life. They are the places where divers are most likely to see sharks.

Barbados also has a large number of shipwrecks, including some of the earliest to be deliberately sunk for divers. Perhaps the best known, and the largest, the SS *Stavronikita*, was purchased by the state and sunk for divers in 1978, while within the Carlisle Bay Marine Park there are at least six wrecks, all in shallow water and accessible to beginner divers. The intervening years have seen a proliferation of life on some of these wrecks – sponges, anemones, and soft coral abound on their outer surfaces, and it is not unusual to see sergeant majors guarding patches of eggs which they have laid on the smooth surfaces of a ship's hull. For the six to eight days until the eggs hatch, male sergeant majors stand guard, sometimes fanning the eggs with their fins, or picking out the dead or imperfect ones. They will also vigorously defend the eggs from would-be predators, often taking on attackers far larger than themselves. Within the wrecks, schools of soldierfish and squirrelfish lurk in the darker recesses, together with the occasional school of glassy sweepers and sometimes a large margate or two. The fish around the Carlisle Bay wrecks are used to being fed and in places dense crowds of sergeant majors and grunts gather around divers. This is also a good place to find seahorses, while scorpionfish and even batfish are sometimes to be seen.

The largest of the Caribbean angelfish, gray angelfish can reach 50 centimeters in length. Large individuals are quite often seen on the deeper reefs off Barbados.

A male sergeant major (lower left of picture, with a much darkened striped flank) is driven back from his clutch of eggs by a marauding horde of blue tangs, rock beauties, and a queen triggerfish.

South America

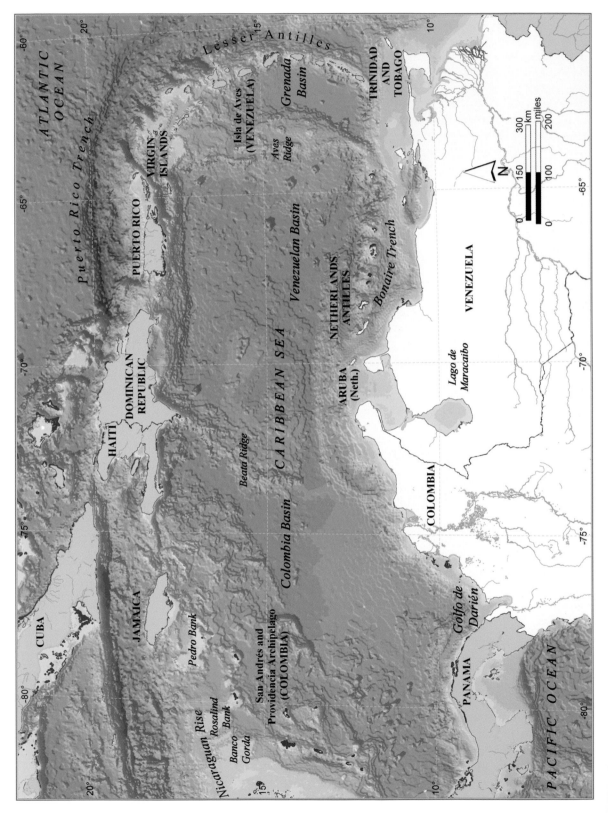

2.5 SOUTH AMERICA

The southern Caribbean Sea is edged with seascapes and landscapes of almost unimaginable variety. Far offshore are remote atolls and archipelagos, unscathed by human change. Closer to land come larger islands, fringed with corals, and then comes the vast bulk of continental South America. Here are extensive mangrove swamps, wet coastal rainforests, rocky cactus deserts, and the highest coastal mountains in the world.

The Caribbean Current sweeps in around the coast of Trinidad and Tobago, and then flows westwards, carrying with it some of the waters and sediments from the great Orinoco River. Above the water, the trade winds blow from the northeast throughout the year. However, this region is spared the ravages of hurricanes, as these great storms are only generated towards the edges of the tropics and this coast is close enough to the Equator to avoid them.

As a general rule coral reefs and large continents do not go well together. Corals are highly sensitive to freshwater and to sediments coming off the land, and continents tend to produce a lot of both. Things are exacerbated along parts of the South American coastline by cool waters, rich in nutrients, which are swept up from the deeper parts of the sea (upwellings) into coastal waters. For these reasons, there are only a few coral reefs along the continental coast, although these can be rich in life and particularly beautiful. The best developed reefs of this region, however, are mostly to be found around offshore islands.

Tobago is held in high regard by divers – home to some of the largest and oldest corals in the Caribbean, as well as densely packed fish faunas. Venezuela's coast is bordered by a string of offshore islands and reef systems, such as the Archipiélago Los Roques and the Archipiélago de las Aves – bathed in clear water and home to some of the healthiest reefs in the Caribbean. The Dutch islands are a continuation of this offshore chain. Here the reefs lie tight up against the coastline, and the minutiae of life has attracted as much attention as the swirling schools of jacks, barracuda, and occasional sharks. Colombia's mainland coast and nearshore islands have scattered fringing reefs, one lying in the shadow of fabulous mountain ranges.

Finally, far off in the Caribbean Sea, both Venezuela and Colombia are host to the most remote coral reefs of the region – Isla de Aves (Venezuela) is little more than a rock, but home to vast populations of nesting green turtles. The San Andrés and Providencia Archipelago (Colombia) includes densely packed tourist centers, but also coral atolls barely explored even by scientists.

Mangroves, both on the continental coastlines and in the clear offshore waters of many islands, are a critical home for many young reef fish.

Many of the reefs in the region lie within parks and reserves, and in a few cases these protected areas extend far inland. As most sources of pollution and sediments come from land – from clearing forests, spraying crops, and releasing sewage – the protection of the adjacent land is a considerable advantage to the corals and fish. The Bonaire Marine Park represents another model for conservation. Here the local people have been fully involved and consulted in the rules and regulations for the park, and the visitors, who must pay an entrance fee, are paying for its upkeep.

Colombia

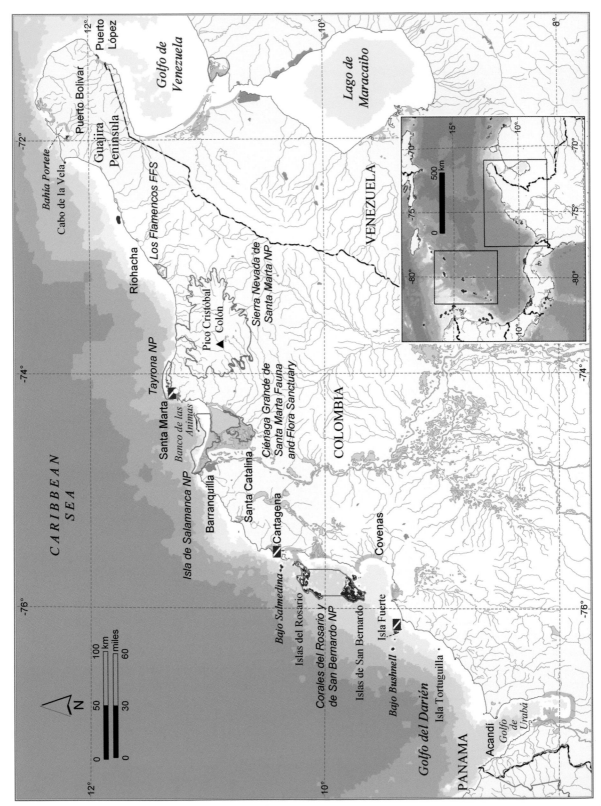

COLOMBIA

Temperature	Santa Marta: 23-31°C (Jan-Feb), 27-31°C (May-Jun)
	San Andrés: 25-28°C (Jan-Mar), 26-29°C (May-Oct)
Rainfall	Santa Marta: 340 mm (rising to more than 2 meters in the Golfo de Urabá)
	San Andrés: 1 910 mm; peak rainfall Jun-Dec
Land area	1 141 957 km^2
Sea area	750 000 km^2
N° of islands	More than 30
The reefs	900 km^2. Most of the reefs are found in the San Andrés and Providencia Archipelago, which includes atolls, platform reefs, and barrier and fringing reefs. Along the mainland there are scattered reefs and coral communities, especially around the offshore islands.
Tourism	There are some 530 000 international tourist arrivals, plus a large number of cruise ship passengers arriving in San Andrés. Diving is popular with locals and tourists, with 11 dive centers on San Andrés and Providencia, and a further 18 on the mainland Caribbean coast.
Conservation	There are growing efforts to protect and manage the reefs. Even the remote oceanic reefs are partly impacted by coral disease and overfishing, although they remain spectacular and important. Many reefs fall within protected areas; however, further strengthening of these areas is needed.

Imagine a place where you look down to see a thriving coral reef, falling away into deep blue waters, then lift your face out of the water to see forested hills rising straight up to snow-capped peaks. Imagine, again, being surrounded by crystal clear waters of blue and shimmering turquoise, with vibrant coral reefs stretching to the far horizon, and being more than 300 kilometers from any inhabited land. Colombia's Caribbean reefs fall into two different worlds.

On the mainland, facing the Caribbean Sea, coral reefs are scattered along a coastline which in places must be considered the most breathtaking in the entire Caribbean. Far out to the northwest, and closer to Nicaragua than to the mainland, a totally different array of oceanic coral reefs and associated islands lies strewn over a huge area, the San Andrés and Providencia Archipelago.

These two worlds are not only physically far apart, but have quite different stories to tell, both above and below water. Mainland Colombia was originally densely populated by Amerindians, but not long after the arrival of Christopher Columbus, in 1499, the Spanish moved in. The first European city, Santa Marta, was founded in 1525, and the rest of the country quickly came under Spanish control, driven by the wealth of gold which was being discovered inland. War, disease, and slavery decimated the native population, but some communities remain, notably in the south and in the coastal mountains. By contrast, the islands of San Andrés and Providencia were uninhabited when first discovered by Spanish sailors in 1510. They were initially

The wide bright lagoon of Isla de San Andrés is a highly productive powerhouse, with rich coral communities as well as seagrasses covering wide expanses.

San Andrés and Providencia Archipelago, Colombia

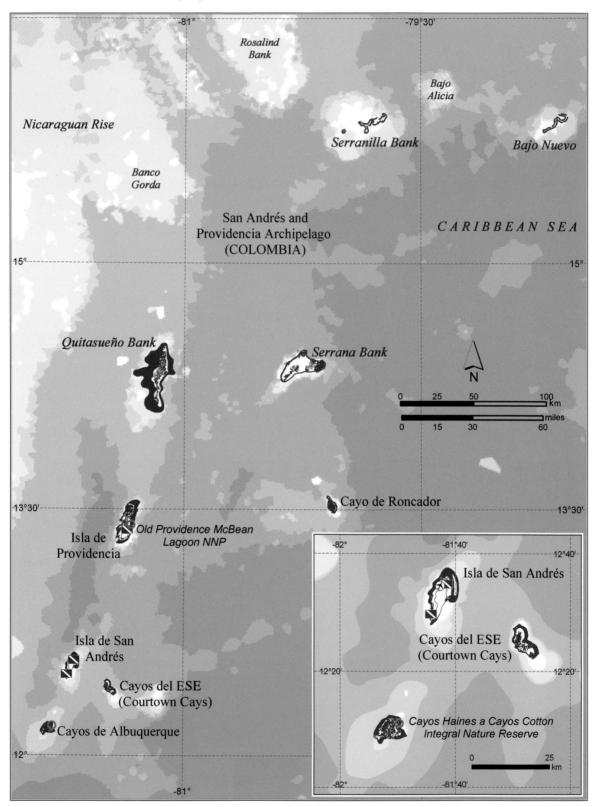

colonized by English and Dutch settlers, and, although they were later ceded to Spain, English remains the first language for many locals.

Marine protection

There is considerable interest in the marine environment in Colombia. The country has, of course, a long history of fishing in most areas, but more recently diving has become a popular pastime, not only for foreign tourists but also for Colombians. This was one of the first Latin American countries to set up a national system to study its coral reefs, and over the coming years this system will help to track changes in the reefs and should support management measures to help restore damaged areas and to reduce threats.

Many of Colombia's reefs lie within national parks, where they have at least partial protection from fishing. Where the parks include adjacent land, they also protect the reefs from sediments and pollution from onshore. Overfishing unfortunately still occurs, including in some of the parks, and larger fish are rarely seen, even around the uninhabited banks and atolls.

The entire San Andrés and Providencia Archipelago, including all of its surrounding waters, was declared a UNESCO biosphere reserve (the Seaflower Biosphere Reserve) in 2000. With a total area of 300 000 square kilometers, this reserve makes up more than 10 percent of the entire Caribbean Sea. Biosphere reserves do not provide strict protection, but encourage a more balanced and coordinated effort to protect the environment in harmony with sustainable practices of human use, such as fishing and tourism. If all the people who depend on coral reefs – fishermen, tour operators, divers, schoolchildren, farmers, and community groups, as well as politicians – can be given a voice in the management of the biosphere reserve then the future for these reefs looks bright indeed.

TOUR OF THE REEFS
Mainland reefs

The coast of mainland Colombia is continuously swept by the northeast trade winds, which are particularly strong in the dry season. These winds run in partial opposition to the Colombian

Streaming over the reef in fast-moving schools, boga are plankton feeders that rely on speed and schooling behavior to outwit their predators.

Countercurrent, a coastal current that spins off the more northerly Caribbean Current and sweeps around to flow back in an easterly direction. The coast is highly varied: dry and rocky in the east, rising to the stunning coastal mountains, then giving way to wetter, lower-lying areas further west.

Eastern Colombia

The Guajira Peninsula in the east is very dry and covered by scrubby desert. Offshore, the waters are home to an important lobster fishery. However, cool upwellings along this coast favor the growth of algae over corals; in some areas there can be extensive development of large seaweeds or macroalgae, particularly during the dry season. Some species can reach lengths of 3-4 meters and one type, called *Cladophyllum*, is found nowhere else in the world.

Vast areas of these macroalgae produce underwater scenes more reminiscent of cold-water coastlines than of the tropical Caribbean. So far only a few coral communities have been found, with the most extensive being in the embayments such as Bahía Portete. For the most part these are low in diversity and have not built true coral reefs. As is typical in areas of upwelling, the offshore waters are highly productive and support significant fisheries, including the capture of sardines and tuna.

Westwards, the coastline undergoes a

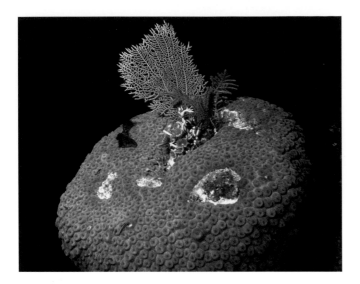

Great star corals are abundant along the mainland coast around Santa Marta. Here dead patches on the colony have been quickly colonized by other corals and fanworms.

remarkable transformation: rocky hills become mountains and the Sierra Nevada de Santa Marta rises up in an array of permanently snow-capped peaks reaching to 5 775 meters. The coastline is wonderful, with deep bays set into the mountain scenery. There are a few mangroves, while underwater there are dramatic walls of rock and coral. Reefs have formed in the more sheltered bays and on western shores which are more protected from the prevailing waves. During the dry season, upwelling water from deeper offshore can again lead to quite cool waters (21°C), while in the wet season runoff from the land can reduce water visibility to 10 meters. Although not so precipitous, the nearby Tayrona National Park is host to green hills, deep bays, and steep underwater scenery.

Many of the reefs and coral communities have grown on steep rocky slopes that plunge to considerable depths. This is a good place to try to distinguish a wide array of stony corals, including the full variety of brain corals, but also branching and elkhorn corals, star corals, lettuce, finger, and fire corals. Although the number of corals has fallen since the 1980s, things have now largely stabilized and may even be beginning to recover.

Barracuda often hang in the deep water off the reefs, just watching divers and snorkelers. Balloonfish are sometimes found floating under overhangs or drifting about at the base of soft

corals. These waters are also great places to see several of the more camouflaged species such as scorpionfish, seahorses, and octopus. In fact, being so well hidden means that these creatures probably see divers far more readily than the divers see them. Efforts to control illegal fishing are beginning to take effect in a few places along this coast, and larger fish such as groupers and snappers are now seen more often.

Western Colombia

Across the wide shallow coastal shelf to the southwest of Cartagena there are a number of coral reefs, mostly around offshore islands or close to the edge of the continental shelf. The Islas del Rosario and adjacent Barú Peninsula are a popular tourist destination. Most of the islands have been developed, with hotels and private dwellings, although there are still some areas of mangroves, and extensive seagrass beds offshore. Fringing reefs are found around the Barú Peninsula, and there are also patch reefs and barrier reefs through the islands. Further south a scattering of uninhabited islands, the Islas de San Bernardo, are surrounded by seagrass beds, shallow sand, and coral-covered plains. All of these reefs and islands fall within a national park, although this offers only partial protection, and overfishing and destructive fishing methods continue to cause problems.

The most dramatic reefs here are the barrier reefs along the outer edges of the seaward islands, where steep walls drop to 35 meters, with sheet corals, some massive corals, and also black coral. In shallower areas there are still occasional patches of elkhorn coral. Fish life is abundant. In shallow water great crowds of sergeant majors can be seen, rising up high above the sea floor in search of food in the plankton. Closer to the corals other damselfish are plentiful. The bright colors of blue chromis are unmistakable, and occasionally divers will see the almost dazzling golden yellow threespot damsel – these are the juveniles, and as they mature their colors fade to a dark gray-brown on top and a pale belly. Often overlooked, threespot damsels are some of the commonest fish on many shallow reefs. They are strongly territorial grazers, defining a small "field" of algae and persistently chasing off other grazers. Also known as farmer fish, these damsels regularly establish new "fields" by nipping at coral polyps, causing them to die and become overgrown with algae.

The small Isla Fuerte is surrounded by a shallow shelf with extensive seagrasses and a few fields of coral, particularly in the northeast. Due west of Isla Fuerte, Bajo Bushnell is a bank which rises up from 60 meters to within 12 meters of the surface. Its northern edge is defined by high ridges and steep slopes or walls, smothered in corals. To the south and east there are wide plains of soft and stony corals, and also considerable numbers of giant barrel sponges. Manta rays occasionally come in over these waters.

The most southerly coastline of the Caribbean Sea enters the Golfo de Urabá, a deep embayment lined with mangrove forests. The inner gulf is too silty for coral growth, but towards the border with Panama there are small reefs. This area is still poorly known, but there is growing tourism, particularly to the small town of Capurgana. An expedition in 1995 described a wealth of corals growing on the rocky coastline and around offshore islets.

San Andrés and Providencia Archipelago

The San Andrés and Providencia Archipelago includes some of the most remote coral reefs in the Caribbean. Spread across a huge expanse of ocean there are only three inhabited islands, plus a few small coral cays on the other shallow reef areas.

San Andrés itself is some 650 kilometers northwest of Colombia's Caribbean coast and 200 kilometers east of Nicaragua. The main population center of the archipelago, San Andrés today has a population of more than 60 000, making it one of the most densely populated islands in the entire Caribbean. The island is built from limestone, deposited millions of years ago as an ancient coral reef and then lifted up by ancient tectonic upheavals. The island of Providencia and its small sister island of Santa Catalina, by contrast, have a population of just 4 500. These islands are quite different: formed by a volcanic intrusion, they are thickly forested and hilly, rising to 360 meters. Tourism is the major industry in San Andrés and is growing fast on Providencia. There are considerable opportunities for diving around the main islands.

Growing in bright clear waters, the reefs

of this archipelago are host to a tremendous diversity of life. More than 270 species of fish from 54 families have been found so far, and 49 species of stony corals from around San Andrés alone. There is also a large range of reef types. Fringing reefs come close to San Andrés and Providencia in a few places, but for the most part these islands are encircled by barrier reefs, separated from land by a deep lagoon. There are four coral atolls in the archipelago: Albuquerque, Courtown, Roncador, and Serrana. These have an outer ring of corals around a central lagoon and each has one or more small coral cays, often teeming with frigate birds, noddies, and boobies. Serrana and Roncador are probably the healthiest of these reefs, with relatively low fishing pressure

Larger predators, such as this yellowfin grouper, are still found on the deeper walls of many reefs. Yellowfin groupers have been recorded a meter in length and weighing over 15 kilos.

and extensive stands of corals. Towards the north, the vast shallow banks of Quitasueño, Serranilla, Bajo Alicia, and Bajo Nuevo are regarded as banks or shoals rather than true atolls, although the difference is subtle and best left for the scientists to argue over.

Despite the varied range of reefs, they have many characteristics in common. The best developed reefs are on the eastern sides, despite the constant pounding by surf. Few true corals survive in the shallowest water, but fire corals abound, along with white encrusting zoanthids. Deep spur and groove formations direct the surge back and forth. Typically there is then a terrace which slopes gently down to about 25 meters, where the sea floor is tightly packed with corals and an incredible abundance of other life. In places there are patches of elkhorn and staghorn corals, and these are particularly extensive on the eastern tip of Serrana, with brain

HAMLETS – SMALL COUSINS OF THE GROUPERS

The diminutive hamlets are a group of fish which look a little like damselfish but are in fact closely related to the groupers. There are at least 11 highly distinctive hamlets, each beautifully, and often quite delicately, colored. Some are widespread; others, such as the masked hamlet which was discovered on Providencia, are found only around a few islands. They are usually seen on their own, hovering or lurking around the base of the corals. Look closely and you can see that

Butter hamlet.

Barred hamlet.

Indigo hamlet.

Shy hamlet.

they have strong jaws like other groupers. They are predators, and spend their days seeking shrimps, crabs, and small fish.

Scientists have been quite confused by the hamlets for many years. Despite their different colors, they are almost identical in their anatomy, and genetic studies have been unable to pick out major differences. Hybrids between different hamlets are quite common, giving further indication of their close relations. The key, and still unanswered, question is whether they are all the same species, with varieties in much the same way that dogs have different breeds, or whether they are different species but still sufficiently closely related to interbreed.

There is also speculation as to why there are so many different types. It has been suggested that certain forms, such as the black hamlet and the blue hamlet, may mimic other fish (the dusky damselfish and the blue chromis in these cases). Resembling a non-predatory fish might enable the hamlets to get closer to their prey.

corals, and fields of sea plumes, sea rods, and branching sea whips. Here, too, is a great variety of sponges and algae.

Below about 25 meters the sea floor drops sharply to form a wall, still richly encrusted with life, including large amounts of sheet or plate corals. Larger fish, including dog snappers, tiger and yellowfin groupers, and jacks are often seen on these deeper slopes. Because the water around these reefs is so clear, light can penetrate to a great depth and coral cover continues well below the limits of most diving.

The lagoons are a mix of sand and rubble in water reaching to 10 or 20 meters deep. There are some wide patch reefs and long, winding, ribbon-shaped reefs made up of brain corals and various branching corals. Occasional pillar corals rise up like castles from the sand. These are places of great beauty for snorkelers and divers, teeming with fish. Elsewhere, in some of the lagoons, wide meadows of seagrass can be found. Many reef fish use the meadows as a breeding and nursery ground, and they are also a popular feeding site for green and loggerhead turtles.

On the leeward, eastern side of the reefs the water typically falls away gradually over a bare rock slope with occasional corals. In slightly deeper water mixed coral gardens can be found in quantity. Large schools of fish such as creole wrasse, boga, and Bermuda chub are often seen swirling in the waters above these reefs. The metallic blue and faintly striped boga often swim in mixed schools with the creole wrasse, feeding on plankton. They are fast, and if threatened tend to flee as a school, rather than seeking shelter on the reef. One group of species is perhaps more abundant on these coral reefs than anywhere else in the Caribbean: the hamlets (see box). Whereas two or three varieties can be seen on almost any Caribbean reef, almost the entire range can be found here.

Although the coral reefs of these islands have received less attention from scientists than many other areas, they have been found to be highly varied. On most reefs, fishing pressure has reduced the numbers of very large fish, but groupers and snappers are still regularly observed on the deeper reef slopes, and nurse sharks are relatively common. Bottlenose dolphins are often seen in the waters around and have been known to come up and investigate divers.

VENEZUELA

Temperature	Barcelona: 21-30°C (Jan), 24-31°C (May-Jun)
Rainfall	Barcelona: 650 mm; wetter from Jun-Aug
Land area	916 560 km^2
Sea area	522 000 km^2
N° of islands	About 100
The reefs	480 km^2. There are scattered coral reefs along parts of the mainland coast, and around the large islands of Margarita and its neighbors, but the largest reefs are around the remote offshore islands. Surrounded by clear, clean waters and unaffected by hurricanes, these reefs are particularly healthy, with a spectacular diversity of life.
Tourism	469 000 visitors. There are 29 dive centers which provide access to many areas, but it is difficult to get to the more remote offshore reefs.
Conservation	Many of Venezuela's reefs lie within protected areas. Some of these are well managed to control fishing and other potentially damaging activities, while remoteness, and a military presence on some islands, deters damage to others.

Venezuela is a vast land, home to an incredible diversity of landscapes: high mountains and plateaus; forested lowlands with a wealth of wildlife; savannas; deserts; mangrove forests; and offshore islands. Given this variety, the offshore waters are often overlooked, but Venezuela is a major coral reef nation.

Far offshore there are extensive reefs, particularly around the las Aves and Los Roques archipelagos. In a few places closer to the mainland, some patch reefs and fringing reefs are found, lying tight up against seagrasses and mangrove forests. Venezuela also has a coastline facing the Atlantic; here there are no coral reefs

Red-footed boobies are masterful flyers, even chasing flying fish in the air.

Mangroves can be fascinating places to explore both above and below water.

Venezuela

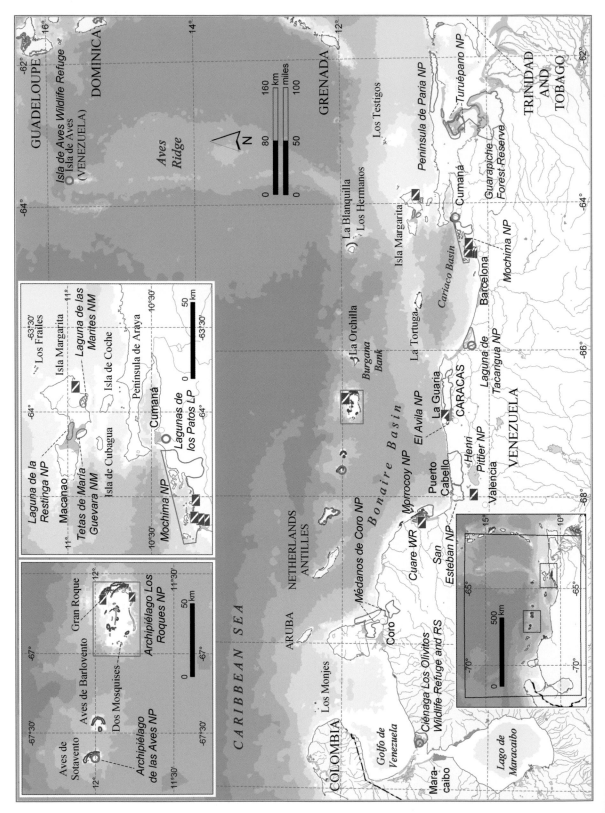

but the vast delta of the Orinoco River fans out through extensive mangrove forests.

Much of Venezuela's economic base is industrial. Oil, particularly from extensive reserves in the lowlands around Lago de Maracaibo, has been a major source of income, and oil tankers ply the shipping lanes all around the country. By contrast both agriculture and fishing make up a fairly minor part of the economy, and large tracts of land and sea remain almost unaltered by humans.

The country has an unrivaled area of national parks and nature reserves. Several of these extend over land and sea, offering protection to the marine resources not only from direct impacts in the water but also from the potentially devastating impacts of land clearance and pollution. The Morrocoy National Park provides a good example of how important such protection can be. In the 1970s tourism to this area was just taking off, driven by bright beaches and warm clear waters. Unfortunately, new buildings began to appear all along the coast and even offshore, with houses being built on stilts directly over the coral reefs and in the shallow lagoons. These reefs would undoubtedly have died, and the water become polluted, had the Morrocoy National Park not been established in 1974. This stunning coastline would have become degraded and its value for tourism lost.

TOUR OF THE REEFS
Mainland reefs

Coral reefs are not widespread along the mainland coast of Venezuela. Their growth is inhibited by freshwater and drifting sediments, exacerbated in some places by cool nutrient-rich upwellings.

In the east, scattered corals are found along the coasts of the Paria and Araya Peninsulas, but west of here upwellings from the Cariaco Basin inhibit their growth. One exception is in the Mochima National Park – a hilly wilderness with extensive rainforest and many hidden bays and beaches. Mochima Bay, a deep inlet in this park, is less affected by the upwellings and contains some quite extensive coral communities. Although not true reefs, these offer some exciting

The massive starlet coral thrives in coastal areas, being more resistant than many other stony corals to sedimentation, here with young parrotfish and bluehead wrasse.

diving, including several shipwrecks. Manta rays and turtles are sometimes seen in these waters.

The best developed mainland reefs are found towards the west in the national parks of Morrocoy and San Esteban. Morrocoy has an extensive shallow shelf with mangroves, small islands, seagrass beds, and both fringing and patch reefs. Most areas are quite shallow, but some reefs in the north descend to 25 meters. Unfortunately a strange sequence of events led to a die-off of some of these reefs in 1996. It is

One of the largest of the parrotfish, the midnight parrotfish reaches over a meter in length.

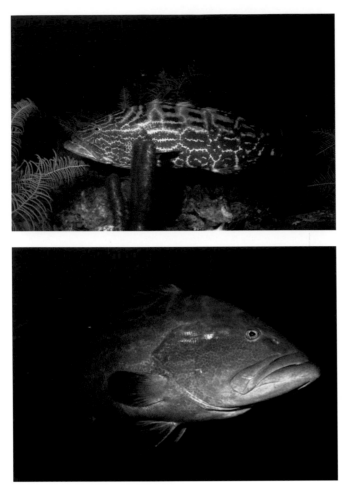

Normally clearly blotched in appearance, the black grouper can change color rapidly to become quite dark.

still a matter of debate, but it seems likely that the cause was an unusual upwelling of nutrient-rich water into the area, which led to a plankton bloom. With calm seas and low winds the plankton sapped the oxygen from the water and many corals, along with fish and other animals, suffocated and died. Some reefs are only just beginning to recover, but others were less badly affected and it is still possible to find some beautiful places where finger and star corals sit amongst fields of sea fans and sea plumes. Yellow goatfish are seen drifting amongst the corals and French angelfish are common. Schools of southern sennet sometimes sweep in over the reefs like a great throng of silver arrows.

The reefs around San Esteban were un-affected by the events of 1996 and, although they are close to the busy sea port of Puerto Cabello, they are home to a number of fringing and patch reefs, as well as seagrasses and mangroves. Star coral and brain corals are common, and there are occasional patches of elkhorn coral. Divers regularly visit this area, where there are a number of shipwrecks dating back to the Second World War.

Offshore islands

Most of Venezuela's coral reefs are found far offshore, around the country's many islands in the Caribbean Sea. Furthest from the mainland is a loose chain of islands, with fringing, bank, and atoll reefs that are among the richest in the Caribbean. The waters here are crystal clear, and unaffected by pollution or sediments from the land.

Archipiélago de las Aves

Not to be confused with the Isla de Aves (see page 214), the westernmost of Venezuela's oceanic islands lie in an archipelago formed by two coral atolls, separated by deep water. Apart from visiting government representatives, the islands are uninhabited. They are low lying, with wide mangrove forests, and many are home to spectacular colonies of red-footed boobies, brown noddies, brown boobies, and magnificent frigate birds. Reefs surround the atolls, with steep slopes dropping into oceanic depths, while smaller patch reefs lie scattered through the lagoons. There have been few visits by scientists to this area, but the reefs are considered to be similar to those of the nearby Los Roques. Turtles are common, and the lagoons are home to large quantities of conch and lobster.

It was long believed that these reefs were the final resting place of part of a large fleet of French vessels which sank in 1678, and in 1997/98 an expedition at last discovered firm evidence of the shipwrecks, with numerous, heavily encrusted cannons scattered amongst the bright reef scenery. About half of the 30 vessel fleet, commanded by Count Jean d'Estrées, went down here, taking 500 of the crew with them. A further 700 crew died of starvation and disease after being stranded on the islands.

Archipiélago Los Roques

To the east of las Aves, Los Roques is Venezuela's largest coral reef system, a fabulous swirl of islands and coral reefs isolated in deep oceanic waters. These reefs have some of the most diverse coral faunas in the Caribbean.

Long barrier reefs run along the eastern and southern sides of Los Roques, with steep outer walls, dropping to 60 meters or more. A deep lagoon runs between these barrier reefs and the islands, while the islands themselves have fringing reefs along their outer shores. Most of the center of Los Roques is shallow, typically about 4 meters deep, with wide seagrass beds. There are numerous patch reefs in the lagoon with dense thickets of staghorn coral and extensive areas of mustard hill coral, boulder star coral, finger coral, and fire coral. The central and northern reefs are surrounded by caves, some providing a home to resting nurse sharks.

In deeper water the reefs are a bewildering tumble of corals. Fish are plentiful everywhere. Perhaps the most numerous are the parrotfish – stoplight, queen, and striped are abundant, and even midnight parrotfish are widespread. Blue tangs and ocean surgeons seem to flow amongst the corals, while densely packed groups of French, bluestriped, and smallmouth grunts drift amongst the scenery. Smaller groupers such as graysbys and coneys are quite common, and there is the occasional tiger grouper.

The larger black grouper is also widespread. Like all groupers these can change color quickly and easily, but despite the name they are rarely black – sometimes dark brown or dark gray, but almost always marked with a pattern of paler lines and darker blotches. Like many big groupers they behave with a strange mix of inquisitiveness and shyness, and when curiosity gets the better of them they will come up to inspect divers. As on many atolls and barrier reefs where there are channels into the lagoon, larger fish such as cubera snapper, jacks, and barracuda are often found patrolling the shifting currents, and Caribbean reef sharks sometimes come in to feed.

Tourism has reached these islands and the small settlement on Gran Roque is now largely focused on this booming industry. The entire island and reef system has been a national park since 1972 and there is now a system of zones: some areas are closed to all fishing, and some are totally closed to visitors. There remains a small commercial fishery but certain restrictions apply,

including no fishing with nets, and a closed season for lobster capture. There are no other settlements, but in the far southwest on the island of Dos Mosquises there is a research station, the Fundación Científica Los Roques.

La Orchilla, La Blanquilla

Two other substantial islands continue the outer island chain of Venezuela. La Orchilla is low lying, with mangroves and small islands nearby. It has been a military base for several decades, although limited access is now permitted. La Blanquilla is also low lying and has a scattering of islets, Los Hermanos, out to the southeast. The waters around both islands remain little known,

---

NEARLY AN ATOLL

Although described by some as an atoll, Los Roques is in fact a "near-atoll". True atolls are formed as corals grow over a sinking volcanic island. Once all of the original island has gone, a near circular formation of corals is left with just a few islands, called coral cays. These islands are very much part of the coral reef system, formed by corals and sand thrown up into piles by waves and currents, particularly during storms (see Chapter 1.1). As the original island sinks, so the corals grow to keep close to the surface and the sunlight. There are some atolls in the Pacific which have been growing like this for 50 million years or more, and one in the Marshall Islands has been measured to a depth of 1.4 kilometers of coral limestone on top of the original volcano.

In the case of Los Roques, this process is not quite complete – one island, Gran Roque, is hilly, and the rocks which make up these hills are actually igneous – formed from molten magma coming from the Earth's core. Until this island itself sinks, over the next few thousand years, Los Roques will not be a true atoll.

Los Roques.

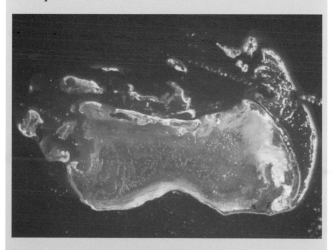

The tiny, remote Isla de Aves is a critical nesting ground for hundreds of green turtles.

but both have extensive fringing reefs, with healthy and diverse coral communities, probably very similar to those of Los Roques. La Blanquilla is occasionally visited by fishers and by tourists staying on Margarita.

Margarita, Cubagua, and Coche

The large island of Margarita is actually two: the western end, the Península de Macanao is hot and dry, and sparsely populated. It is connected to the larger eastern part by a narrow sandspit, behind which lies the vast Laguna de la Restinga, a wetland wilderness with mangrove forests and shallow waters that are home to a fascinating range of life, including turtles and flamingos. The whole island is a mix of wilderness and urbanization. There has been extensive, even high-rise, development in places, thanks to the tourist trade and to the island's free port status. To the south there are two other large, low-lying islands, Cubagua and Coche.

All of these islands, and particularly Coche, were once home to a major pearl industry, and pearl oysters are still found in abundance in places. The oysters feed by filtering the water for the plentiful nutrients which are found here, encouraged by cool upwellings from the Cariaco Basin to the south. But these same conditions mean that the water is sometimes quite murky and there are only scattered patch reefs around the islands. These are home to some interesting fish communities, including large schools of grunts, but also more unusual species – lesser electric rays and the vieja, a medium-sized, mottled orange-brown grouper found only in the waters off South America. To the northeast of Margarita, Los Frailes is a group of high rocky islands with clearer waters, popular amongst divers and snorkelers coming on day trips from Margarita.

In addition to these main islands, there are two other small island groups off Venezuela's coast. In the east, Los Testigos, or "the witnesses", are a group of three high islands with a small resident fishing population. The islands lie in the path of the waters swept up from the Orinoco River, so strong currents and poor visibility are typical. There are no true reefs: a few brain and fan corals survive here, but for the most part low-profile encrusting sponges and tunicates smother the surfaces. However, with the abundant nutrients and plankton, fish life in these waters is prolific.

At the other end of the country another small group of high rocky islands, Los Monjes, lies near the mouth of the Golfo de Venezuela. These islands, with their high cliffs, are an important nesting ground for birds such as brown boobies. Some have well developed coral communities, but these remain little known and largely inaccessible.

Isla de Aves

Far, far away, the Isla de Aves lies some 700 kilometers north of Venezuela, and about 200 kilometers from its nearest neighbors in the northern Lesser Antilles. Looking almost marooned in deep waters, this tiny islet is only 400 meters long. There is a small military and scientific base, but the island is predominantly a home for wildlife. Rich coral reefs are found on all sides, dropping precipitously to the north and east, but with a few other bank reefs rising in the nearby waters to the south. The island is a critical nesting ground for thousands of green turtles, so it is particularly unfortunate that it may soon disappear. During Hurricane David in 1979 the island was completely submerged for about eight hours, and the pounding waves greatly reduced its total area, as well as removing some 50 000 turtle eggs that had been laid in its sandy beaches.

ARUBA

Temperature	24-29°C (Jan-Feb), 26-31°C (May-Oct)
Rainfall	510 mm
Land area	183 km²
Sea area	6 000
N° of islands	5 (2 small)
The reefs	Less than 50 km². Rich coral fields are found on a gently shelving coastal plain.
Tourism	691 400 visitors, plus 487 300 cruise ship arrivals. Ten dive centers – diving is only one of many attractions for tourists.
Conservation	There are no marine protected areas. Mooring buoys protect the reefs from anchor damage.

Aruba has arid, rocky terrain rising to 189 meters. It is a Dutch island run as a separate political entity from the neighboring Netherlands Antilles. Although claimed for Spain in 1499, settlement was limited, and native Arawak peoples survived here long after most other islands. Dutch is the official language although most people speak Papiamento, a creole language based on Spanish, Portuguese, and Dutch – perhaps a good reflection of the island's diverse origins.

Tourism is now the main economic activity and, although dive tourism is only a small element, the island is known for its wreck diving. The freighter *Antilla*, at over 120 meters long, is one of the largest wrecks in the Caribbean. She was scuttled here in 1940 (despite being newly built) following the German invasion of the Netherlands. Lying in very shallow water this wreck offers good opportunities for more adventurous snorkeling as well as for diving.

Aruba lies on the South American continental shelf, only about 30 kilometers from Venezuela. As a result, the underwater scenery mostly consists of gentle slopes, and in a few places it is possible to swim out to the coral reefs directly from the shore.

In the shallow waters there are some wide and beautiful stands of elkhorn coral. Further offshore, large areas are dominated by soft corals, with occasional pillar corals, as well as some brain and star corals. The deeper waters and steeper slopes have more sheet corals and star corals, with some deepwater sea fans. Most of the common reef fish are found here, including gently moving butterflyfish and larger angelfish. Looking close amongst this scenery one small damsel, the sunshinefish, is abundant – the adults are a somewhat dull, dark brownish gray, easily overlooked. By contrast, the juveniles, often seen in small groups, seem to almost glow, with their bright yellow backs and bluish purple bellies.

The yellowtail snapper is found in abundance in these waters, as in so many places across the Caribbean. These fish are opportunists, mostly feeding on plankton but often attracted to divers by the chance of free handouts. It has been suggested that their color pattern may be a direct mimic of the yellow goatfish, and indeed they are often seen swimming amongst schools of these fish, perhaps using them as a cover to pass unseen and thereby capture small fish that swim close by.

Towards the south the waters drop down to 40 meters. Here and in the far north there can be stronger currents and more swell on the surface. Barracuda and larger jacks venture into these waters, and spotted eagle rays make the occasional breathtaking fly-past.

The yellowtail snapper is an opportunist, and a familiar companion to divers and snorkelers across the Caribbean.

Aruba and the Netherlands Antilles

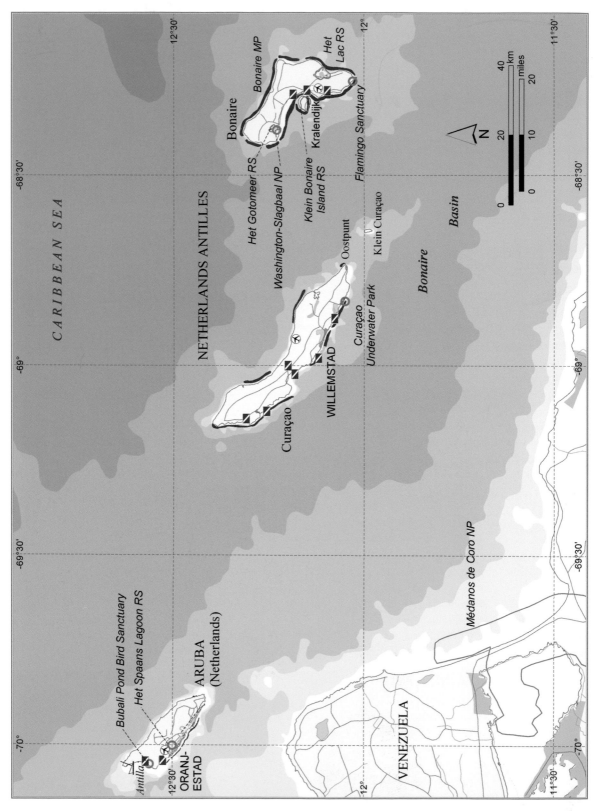

NETHERLANDS ANTILLES
BONAIRE AND CURAÇAO

Temperature	Willemstad: 24-28°C (Jan), 24-31°C (May-Oct)
Rainfall	540 mm; slightly wetter from Oct-Dec
Land area	732 km^2
N° of islands	4
The reefs	There are fringing reefs all around both islands, in places coming within 20 meters of the coast, offering superb opportunities for snorkeling and shore diving. Bonaire's reefs are in a particularly healthy state, with extensive corals and teeming fish.
Tourism	255 000 visitors, plus 340 600 cruise ship passengers. There are 25 dive centers on Curaçao and 15 on Bonaire. Dive tourism is one of the major attractions on Bonaire and is rapidly expanding in Curaçao.
Conservation	The Bonaire Marine Park is widely regarded as a model for sustainable conservation: divers, snorkelers, and boat owners pay a visitors' fee which directly funds park management. Efforts to protect the reefs have been less effective in Curaçao, but may now be improving.

The small islands of Bonaire and Curaçao are encircled by vibrant coral reefs, and have become synonymous with fabulous diving and snorkeling, and tightly packed marine life. There is little runoff from the land and the islands are surrounded by deep ocean, so the water is nearly always crystal clear. Fringing reefs run close to the shore on all sides of the islands. There are opportunities to see larger creatures, but many who come here are entranced by the fine details, the delicate movements of cleaner shrimps in the arms of sea anemones, or the darting jewels of blue chromis swimming amongst corals and sponges.

The Spanish came across the islands of Bonaire, Curaçao, and Aruba in 1499, and used them sporadically for rearing livestock or holding prisoners. In the early 17th century the Dutch took over the islands. Curaçao became one the major trading centers for slaves in the Caribbean, while Bonaire became an important salt producer, an activity which continues today.

Aruba (see page 215) is now politically separate, while Bonaire and Curaçao remain part of the Netherlands Antilles, along with Saba, St Eustatius, and St Maarten in the Lesser Antilles (Chapter 2.4). Together they form an autonomous part of the Kingdom of the Netherlands.

Willemstad, in Curaçao, is the capital of the Netherlands Antilles. Adjacent to the sizable harbor behind Willemstad is one of the largest oil refineries in the world. While this is an important source of revenue for the island, the white sand beaches and warm clear waters have

The diminutive harlequin bass is easily overlooked. These fish form close partnerships, with a male and female patrolling a joint territory and hunting together.

now become a major draw to international tourism. For Bonaire tourism is far and away the most important industry.

Bonaire also stands proud as one of the best examples in the Caribbean of how to conduct marine conservation. The Bonaire Marine Park surrounds the entire island down to a depth of

For photographers, the fine details of life in the Netherlands Antilles are a great draw. Here the large tentacles of a giant anemone are set against an encrusting red sponge.

60 meters. There are strict restrictions on fishing, and some areas are fully closed. Divers may not touch or collect anything, and, where boats are used, they must use mooring facilities. To help run the park, users pay a fee. This goes towards the employment of rangers who undertake a variety of activities from education, to scientific research, to patrolling the reefs.

In contrast, Curaçao has localized pollution problems and overfishing is quite widespread. The Curaçao Underwater Park stretches for 21 kilometers along the southernmost coast of the island out to Oostpunt (East Point), but no parts are fully protected from fishing. Attitudes are changing, however – there are growing efforts to enforce a 25 year old ban on spear fishing, and there is also talk about a second marine park for the northwest.

ISLAND TOUR
Bonaire

The fringing reef begins very close to the coast all along the leeward (western) shore, and most people get to the reef simply by walking and swimming directly from the beach. It is important that those who reach the reef in this way make every effort to walk only on sand or marked paths; stepping on corals and other marine life quickly kills them.

In most places the reef starts with gently sloping gardens of soft corals. Although much rarer than it once was, there are still some areas of thriving elkhorn coral – a particular delight for snorkelers, who can spend hours peering ever deeper into their complex branching depths.

Further offshore the reef slopes downwards, sometimes with steep inclines. Giant colonies of star and brain corals are found amidst a seascape of wafting sea rods and plumes, and wider sheets of sea fans. Many of the usual sponges are widespread – elephant ear, rope sponges, and some really spectacular stovepipe sponges whose long tubes can reach over a meter and a half in length. In deeper water, sheet and scroll corals are abundant, and wire corals poke out from the reef slope. The devil's sea whip is also found in deeper waters – these long, pencil thin strands, bedecked with polyps, are gorgonian soft corals that can reach over 2 meters in length.

Bonaire's reefs are host to a constantly changing and lively array of Caribbean reef fish. Yellowtail snappers are abundant, and dense schools of schoolmasters and mahogany snappers drift alongside schools of bluestriped grunts. Graysbys are the commonest of the groupers, but there are also many tiger groupers, and the yellowmouth grouper can often be seen. The last are rather plain groupers, usually a pale brown in color. They are often seen at cleaner stations where gobies come in to remove parasites; to signal their desire for these services, yellowmouth groupers switch to become almost black.

The smaller life of Bonaire's reefs has thrilled many underwater photographers. Seahorses and frogfish, though rare, are occasionally found. Other species which are common on many Caribbean reefs can be easily admired in these clear, calm waters. Harlequin bass and tobacco-fish, for example, are small predators found on most reefs. They are close relatives of the much larger groupers, but are more elongate and elaborately patterned – they tend to feed on small fish and crustaceans. Harlequin bass are particularly interesting: they form tightly bonded mating pairs that patrol a joint territory and hunt together in a highly effective manner, approaching an unsuspecting prey from opposite directions so that it flees from one straight into the mouth of the other.

Towards the south of the island there is a stretch of "double reef": the first reef slope drops down to a sandy channel at 18-25 meters, beyond which a second reef rises up again. The channel provides a stunning blue backdrop to the reefs, but is also a great place to spot a range of sea creatures. Southern stingrays are not unusual, and there are many areas where garden eels can be seen, shifting slowly in the gently moving waters. Blue parrotfish may venture across the sand to pick at algae. Another common sight is the sand tilefish, a cream-colored, eel-like fish which swims and hovers in the water with very graceful undulations of its elongated dorsal and anal fins. Females maintain tunnels in the sand and mark the area with wide patches of stones, making it harder for predators to dig down to the tunnels.

In the far north and south of Bonaire, and indeed in the few areas where divers venture eastwards, currents can be quite strong and unpredictable. Here the surging waves have sculpted deep spur and groove structures into the shallow reef areas. Sharks and eagle rays are occasionally seen here.

Bonaire is an exciting island for wildlife on land, too. There is a large national park in the north which, dominated by dry forest, is home to parrots, hummingbirds, and giant iguanas.

There are also shallow coastal swamps and lagoons with hypersaline areas and mangrove forests. Common in these parts are pelicans and a variety of egrets and herons, as well as many smaller wading birds. Flamingos favor areas

A TIME TO BREED

The timing of coral spawning on these islands is now well documented and this has become a significant attraction for divers. Coral spawning (see page 21) is linked to phases of the moon, and in Bonaire and Curaçao it always takes place after the full moons of September and October.

In fact it is not just corals that choose to spawn at this time; many other reef inhabitants do the same, and observant divers may easily spot unusual behavior from a host of creatures. From midday until early evening sea urchins, sea cucumbers, and even some sponges may be seen to be spawning – in each case releasing small clouds of faint white particles into the water. Later in the evening others join the show. Brittle stars clamber to a high point on the reef surface to broadcast eggs and sperm into the surrounding waters.

But it is the corals that are the most impressive. Four days after the full moon, between 9.15 and 9.45 pm, in an unerringly well timed event, all of the elkhorn corals release their spawn into the waters. The following night they may give a second performance, but that next night is largely given over to the staghorn corals, and the night after to the star corals.

It seems so incredible that these tiny animals, with no brains, no means of locomotion, and no eyes, can work in such synchrony. A diver floating amongst the corals can watch, entranced, as the apparently inanimate form of a large coral colony suddenly releases a great cloud of spawn. All the more amazing is when the diver looks around and sees that all the other corals of the same species are doing the same thing at exactly the same time.

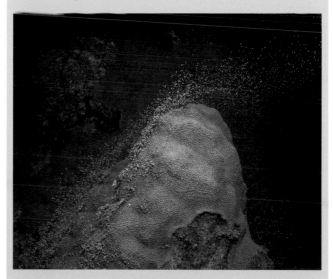

Blue parrotfish, one of the largest of the parrotfish in the Caribbean, often venture over the deep sandy areas to graze on the fine algae growing on the sand's surface.

where salt concentrations are particularly high and where they can filter the water for flies and small crustaceans. The number of flamingos varies enormously from year to year, but it can reach the thousands. The importance of these areas has been recognized and several have been listed as Ramsar Sites: places of global importance for their wetlands and wild bird species.

Curaçao

As on Bonaire, the reefs around Curaçao begin 20-250 meters from the coast, easily accessible to divers and snorkelers without the need for a boat. An abundance of soft corals descends in a gently sloping plain to about 7-12 meters' depth. After this a steep drop-off plunges down into rich blue vistas towards a sandy base at 50-90 meters. Boulder star corals are widespread, in places forming a patchwork of high sculpted mounds over wide areas. Yellow finger corals, sheet corals, and scroll corals are also common.

Large numbers of French grunts and tomtates rest out the day in the shallow reef, and closer amongst the corals the fine form of the juvenile spotted drum brightens up many darker recesses. Butterflyfish are abundant, the foureye butterflyfish particularly so. This butterflyfish is one of the few direct predators of corals, regularly taking a nip at individual polyps, notably of soft corals, despite the fact that they have quite a broad diet. Their impact on the corals as a whole is probably negligible.

Sandy patches in the reef are a good place to look out for the well hidden form of the sand diver, a small predator that sits, often half buried in the sand, waiting for unsuspecting prey. Goatfish may also sometimes be seen, using the strange barbels on their chins to wriggle in the sand for food.

The healthiest reefs are found in the far north and south of the island. Around the northern coastline there are wild rocky shores with sharp rocks, cliffs, and small caves. These areas can be quite exposed to waves and currents. Nurse sharks are regularly seen resting on the sea floor, and hammerhead and whitetip sharks occasionally venture into these waters.

Around the capital there have been declines in corals and fish, associated with pollution and with sediments produced by coastal construction, but further south the reef life picks up again. There are various species of moray eels, including the large green morays, often seen foraging amongst the nooks and crannies of the reefs. The chain moray is more usually found resting during the day, with only a head pointing out of a hole – a beautiful creature dappled and striped with bright yellow against a background of dark brown or black.

The small island of Klein Curaçao lies some 15 kilometers to the southeast of Curaçao, and is occasionally visited by divers. With stronger currents and sometimes rough seas, this can offer exciting diving for the more experienced. It is a good place to see larger pelagic fish, while manta rays occasionally come around the reefs.

TRINIDAD AND TOBAGO

Temperature	Canaan (Tobago): 23-28°C (Jan-Feb), 26-30°C (May)
Rainfall	Canaan (Tobago): 590 mm; wetter from Jul-Dec
Land area	5 152 km^2
Sea area	74 000 km^2
N° of islands	34 (7 large)
The reefs	Less than 100 km^2. Both islands are strongly influenced by the sediments and nutrients washed past from the Orinoco River. Tobago has thriving marine communities and some reefs, famed for dramatic scenery, large fish, and manta rays. Trinidad is largely surrounded by quite murky waters, but there are still a few areas of coral.
Tourism	383 100 visitors, plus 82 300 cruise ship arrivals. There are 19 dive centers, all but two on Tobago.
Conservation	The marine environment remains largely unprotected, with the exception of the Buccoo Reef Marine Park; however there is now growing concern for the marine environment on Tobago.

Although sometimes counted as part of the Lesser Antilles, the islands of Trinidad and Tobago actually lie on the South American continental shelf, and their geological roots place them quite apart from the other Caribbean islands to the north. The same South American links are reflected in the islands' wildlife: in the forests of Trinidad it is possible to see howler monkeys, toucans, parrots, and armadillos.

The sea around Trinidad is heavily influenced by the Orinoco River, which enters the Atlantic Ocean to the south, bringing with it 150 million tons of sediments every year. Much of this is swept northwards by the Guyana Current, preventing the growth of corals apart from in a few areas in the north. Tobago lies further offshore, surrounded by deeper, clearer water, and has rich coral reefs along its northern shore.

Trinidad is a large cosmopolitan island; it has a strong industrial base, and is famed for having the region's biggest carnival. The most impressive physical feature of the island is the Northern Range, a ridge of forested mountains running along the northern coast – the rest of the island is a mix of hills and lowlands, with wide areas of agriculture and some further forests. Tobago feels quite different, a small island where tourism and fishing are the dominant activities, and where forested hills fall away to white sand bays.

Nature conservation on land has a long history. The Main Ridge Forest Reserve on Tobago is among the oldest protected rainforests in the world, first designated in 1776, and the island of Little Tobago was protected in 1928. In Trinidad perhaps the best known nature reserve is the Caroni Swamp, a mangrove wilderness where vast flocks of Trinidad's national bird, the scarlet ibis, come to roost every evening at sunset. Unfortunately, efforts to protect the marine environment have been limited, with the

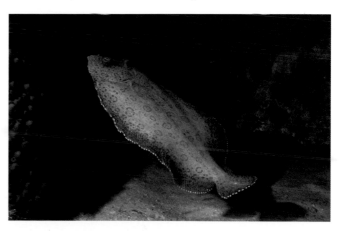

A peacock flounder breaks cover as it flies up from a patch of sand.

exception of Buccoo Reef in Tobago. Problems, ranging from pollution to overfishing, are only just being recognized, but there are growing calls for improvements in reef protection.

ISLAND TOUR
Tobago

At the eastern end of Tobago, the rocky coastline around the small town of Speyside and the island of Little Tobago mark the edge of some wild underwater scenery with steep walls, vast rocks, and fissures. Currents sweep through this area, and some sites are only accessible to more experienced divers. Similar scenes surround the St Giles Islands which are also host to some of the most important bird nesting colonies in the southern Caribbean. Magnificent frigate birds and red-footed boobies, as well as brown boobies, noddies, red-billed tropicbirds, and

Audubon's shearwaters, nest on the islands and soar over the waters all around.

Corals and sponges dominate the scenery of this part of the island in almost equal measure. There are some truly enormous boulder brain corals (see box), set amongst landscapes of sea plumes and rods, barrel sponges, rope sponges, yellow tube sponges, and the occasional brilliant flash of an azure vase sponge. Thanks to a rich supply of nutrients, filter feeders do well here: deepwater sea fans abound on the steeper slopes, while rocky surfaces are host to a great variety of hydroids, tunicates, fanworms, and Christmas tree worms.

Plankton feeding fish are found in almost continuous swirling masses in many areas, including chromis, sergeant majors, and creole wrasse. This is a great place for seeing manta rays. These huge rays appear unafraid of divers, and sometimes seem to enjoy swimming through divers' bubbles (this may help dislodge parasites – they have even been known to rub

Trinidad and Tobago

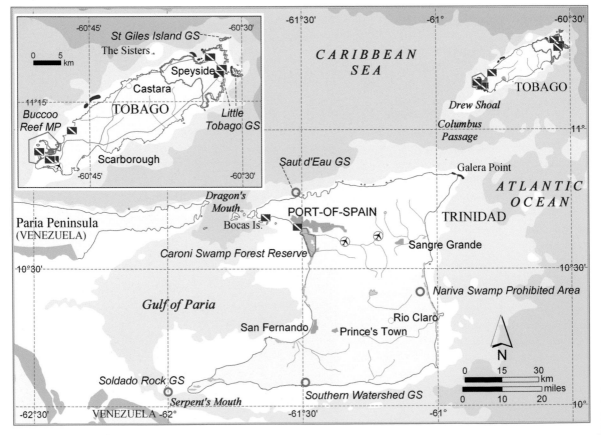

directly up against divers). Mantas give birth to live young which are 1 meter across when born; fully grown individuals have been recorded at 7 meters from wing tip to wing tip.

Also common in the fish-rich waters of eastern Tobago are large numbers of fish-eating fish, or piscivores. Tiger groupers are found close to the bottom, lurking amongst the corals or in the mouths of caves. Schools of tarpon swim past, barracuda are abundant, and there is a great diversity of jacks, including horse-eye, crevalle, and black jacks, and rainbow runners. Black jacks are one of the more unusual species in the Caribbean: they have steep, slightly indented foreheads, and a habit of coming up close to check out divers. Further out in the blue waters, hammerhead sharks are not infrequent visitors.

Along the northern shore of Tobago some of the bays offer opportunities for snorkeling. This coast is calmer, with few currents and some well developed reefs. These are home to the red-spotted hawkfish, a small and wary predator which rests on the bottom, occasionally lunging forwards to capture small crustaceans. White-spotted and green morays are fairly widespread, and in a few places divers may spot the rare comb grouper, brownish gray in color, with paler blotches and very small, white spots.

Looking close up, flamingo tongue snails are often found at the base of sea fans, and these reefs are also home to their much rarer relative, the fingerprint cyphoma – these are more elongate snails, but with the same wide belt around the middle of their shells. Their mantles are marked with fine transverse stripes of dark brown and gold. Both species move up at night to feed on sea fans and other soft corals.

On sandy areas the highly camouflaged peacock flounder is quite common. Flounders are one of a large group of flatfishes which undergo an extraordinary metamorphosis early in life. They begin life in the plankton looking just like any other fish, but after a few weeks a strange change takes place. Their entire body – muscles, skeleton, and blood vessels – begins to twist and flatten out. At the same time, one eye begins to migrate around to the same side of the body as the other. Thus transformed these pancake-like fish are perfectly formed to lie flat on the sand. Close inspection reveals their strangely twisted mouth and eyes, showing that these are actually fish lying on their sides, not their bellies.

In the far west of Tobago, Buccoo Reef Marine Park has been protected since 1973. This area presents a microcosm of Caribbean ecosystems, with mangrove forests along part of the shore, then a wide lagoon with seagrass beds and patch reefs, and bounded by a long, protective coral reef. The offshore reef descends

ANCIENT BRAINS

Massive, dome-shaped corals are the slowest growing of all stony corals and may live to quite an astonishing age. Some may wax and wane with the years – patches may die, but other parts survive and continue to grow up in a process known as fission, sometimes creating complex sculptures with new corals or sponges growing over the dead parts.

Occasionally, however, conditions remain just right, year after year, decade after decade, and century after century, and it would seem that the northeast coast of Tobago is one such place. Here there are quite a large number of boulder brain corals reaching to 2 meters or more in width, and one celebrated giant which measures 6 meters across and 4 meters high. This is one of the largest known corals in the Caribbean, and is likely to be very ancient indeed. Assuming a growth rate of 6-12 millimeters per year, this coral could be between 330 and 660 years old.

It is also interesting to consider how old an individual polyp might be. On some corals, the individual polyps may live for only five to ten years, before they are overgrown by other polyps of the same colony. On some smooth-surfaced brain and star corals, however, the polyps grow outwards in the same direction as the colony itself. If such a colony is split open the individual polyps can be traced back along a narrow tube, like a straw, with the living polyp growing on the outer tip. In these corals it is entirely possible that a small number of the polyps may be as old as the coral itself.

in a beautiful coral-covered slope to depths of around 30 meters. Gray angelfish are quite common, and it is not unusual to see juveniles, strikingly marked in black, traversed by bright golden yellow bars. Within the calm lagoon, the patch reefs are excellent for snorkeling – stony corals, sea fans, and other soft corals rise up out of dazzling turquoise waters. Fish are abundant, with sergeant majors, bluehead wrasse, small

Not all corals have an easy life. Large parts of a colony may die, as has happened with this grooved brain coral, but other parts continue to grow upwards in smaller colonies around the edge.

Thanks to a rich nutrient supply, Tobago's waters are a great place for many filter-feeding organisms, such as these social feather duster worms and yellow tube sponges.

parrotfish, and surgeonfish hovering and darting in all directions.

The adjacent seagrass is also worth a look. These underwater plants are in fact the only true plants to grow completely submerged in the sea, and they often play host to juvenile fish. It is important in these shallow waters not to walk on the bottom except in areas of bare sand. Unfortunately, parts of Buccoo Reef have been damaged by trampling feet, and even today tourists are offered plastic shoes by some tour guides, encouraging them to crush underfoot the very life they have come to see.

Running southwest from Tobago into the Columbus Passage there is a shallow ridge known as Drew Shoal which is swept almost constantly by the Guyana Current. Although there are few corals growing here, there are numerous encrusting and elephant ear sponges, and many large barrel sponges, twisted into strange shapes by the current. French and gray angelfish, which feed on sponges, are abundant. Lurking close to the bottom amongst some complex cuts and overhangs are green morays, while nurse sharks are also common in this area. From June to November sediments swept up from the Orinoco River can occasionally reduce visibility to almost zero.

Trinidad

On Trinidad the clearest waters are along the north coast, but even these are affected by the freshwater running off the mountains – corals grow around a few offshore islets. Some of the most interesting marine life is to be found around the Bocas Islands and the Dragon's Mouth in the northwest. The powerful Guyana Current is channeled through this area before passing into the Caribbean Sea, and whale sharks and manta rays have been seen here, feeding on the abundant plankton.

There are no reefs, but in a few places soft corals and sponges thrive, and large angelfish are abundant. Telestos, the small corals more often seen on wrecks and harbor walls, are common. It is not unusual to see some quite large schools of fish, including porkfish, Bermuda chub, caesar grunts, and jacks, and there is a great diversity of blennies and gobies. Jewfish are found in a few caves, and scorpionfish are sometimes spotted resting on the sea floor.

3.1 REEF SPECIES

The abundance and diversity of life on the reef can seem overwhelming. At the same time, it can be deeply rewarding to try and learn a few names of fish, corals, and other creatures, and there are several excellent guides (see page 248) to help. Here we provide a brief introduction to some of the main groups, followed by a series of images and short descriptions of a few of the most common or interesting species to be found on and around Caribbean reefs. Remember that these are not the only species listed or described in this book; many others are to be found throughout, and the Index is a useful guide.

After the plants (which form a separate "kingdom" in the classification of life), the groupings of animals listed below represent different animal "phyla". A single phylum, or life-form, may contain many thousands of different species, usually sharing some common pattern in their body-plan or structure. Coral reefs are hot-beds of evolution and almost all of the known animal phyla are found here, though many are too small, or too well hidden, to be easily seen.

Throughout this book we have used the most widely known common names for the various plants and animals. Readers should be aware, however, that common names do vary from place to place, and from book to book, so in this chapter we also give the scientific names, which are fixed and do not change.

PLANTS   *page 227*

All reefs depend on plants which lie at the base of the food chain, but many are unseen – microscopic species hidden in the plankton or within the tissues of corals. The simplest visible plants on the reef are the algae – some are little more than a fuzz of green or brown on the bare reef surfaces, while others have formed larger bodies and are called macroalgae. Higher plants – species with a complex vascular system like those which dominate on land – are rare in the sea, but two groups, mangroves and seagrasses, are important parts of tropical coastal ecosystems and are often found adjacent to coral reefs.

A damselfish guards its algal farm, with stony corals all around.

Bright cup corals, black corals, and sponges are found in deeper waters.

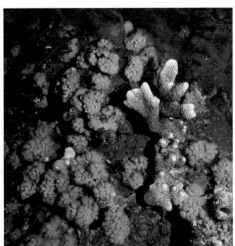

SPONGES: PORIFERA *page 227*

Sponges are the simplest group of multicellular animals. They have no nervous system or means of locomotion, but build simple bodies characterized by numerous open channels. Water is drawn in through fine pores, and leaves through larger pores, having been filtered for food. Many contain complex toxins to deter predators.

CORALS AND THEIR RELATIVES: CNIDARIA *page 228*

A great range of animals make up this group, including jellyfish, sea anemones, and corals. Many have polyp-type bodies (see Chapter 1.1) and many build large colonies. The soft corals, with more flexible skeletons, include a group known as gorgonians, which are the familiar sea fans, sea whips, sea rods, and sea plumes.

The stony corals, belonging to the group, or order, Scleractinia, are the major reef builders, with 62 species found across the Caribbean. They lay down skeletons of calcium carbonate, or limestone. Most species are colonial, and build large structures, although some, such as cup corals, live as solitary polyps. They are sometimes described based on their growth forms (branching, massive mound, boulder, encrusting, sheet).

SEGMENTED WORMS: POLYCHAETA *page 230*

A great host of worm-shaped creatures inhabits the reef, but many pass unseen, burrowing in sand and rock. The most familiar are the feather duster worms, which have a spectacular feathery mouthpart used to filter the plankton for food. Another group includes the free-ranging predatory fire-

A coney passes by a rock beauty.

worms, which feed on corals and are brightly colored, with fine tufts of bristles on their sides.

CRUSTACEANS: CRUSTACEA *page 230*

One of the most abundant and diverse groups on the reef – they are typified by having a tough exterior or exoskeleton and two pairs of antennae, but have diversified into an extraordinary range. Tiny copepods, mysid shrimps, and others swim in the plankton; amphipods can sometimes be seen as parasites on fish, but the decapods, including crabs, hermit crabs, shrimps, and lobsters, are the most familiar.

MOLLUSKS: MOLLUSCA *page 231*

Mollusks are another diverse and important group. Best known are the snails (gastropods), often thought of as having coiled shells, although many on the coral reef have lost their shells altogether. The bivalves are less obvious – these are the scallops and oysters, with a pair of shells hinged at one end. Strangest of all, in many ways, are the squids and octopuses (cephalopods) – they are very advanced animals with large brains and excellent eyesight. Their feet have developed into elongated tentacles, and they are highly mobile.

ECHINODERMS: ECHINODERMATA *page 231*

At first glance the various members of this group seem to have little in common, but they all have a five-pointed symmetry to their bodies (most obvious in the five arms of a starfish). Most also have internal skeletons built from small calcium carbonate plates, and on the outside of their bodies they have tiny tube feet. There are about 150 species in the Caribbean.

VERTEBRATES: VERTEBRATA *page 232*

The creatures with a backbone and a braincase, or cranium, are perhaps the most familiar to us. They include fishes, amphibians, reptiles, birds, and mammals. On reefs, fishes predominate. Reptiles are represented by marine turtles, while a few birds and mammals are regular visitors.

Some 1 400 different fish have been described from the western Atlantic: many hundreds of these make their home on reefs, while hundreds more are occasional visitors. Here we provide descriptions of just a few of the commonest species and groups.

PLANTS

Hanging vine algae Halimeda sp
The green algae are probably the most obvious species on the reefs. Hanging vine algae and their relations are widespread. They lay down calcium carbonate (limestone) in their leaves to make them impalatable to many grazers. Their skeletal remains go on to be important parts of the reef sand.

Y branched algae Dictyota sp
There are three common groupings of algae – brown, green, and red – seen on the reefs. This Dictyota is actually one of the brown algae, although beautifully colored with shades of green or blue.

Manatee grass Syringodium filiforme
Seagrasses are higher plants that have evolved to live in saltwater. Although they look like grass they come from various different plant families. Manatee grass is characterized by its fine blades, cylindrical in cross section.

Turtlegrass Thalassia testudinum
The broad leafy blades of turtlegrass are unmistakable – they often form dense meadows and, in clear water may be found to depths of over 20 meters.

Red mangrove Rhizophora mangle
Mangroves are trees which have adapted to live between the tides. They come from a number of plant families, but each has adapted in special ways. The arching prop roots of the red mangrove help to support the plant in thin or soft soil, but also allow air to get to the roots in the waterlogged environment.

Black mangrove Avicennia germinans
The fine pencil-like roots of the black mangrove are unmistakable. In the waterlogged soils these "pneumato-phores" act as snorkel tubes, allowing the plant to get air into its roots at low tide.

SPONGES: PORIFERA

Lavender rope sponge
Various sponges throw out beautiful rope-like forms, and can range from red to yellow to purple to green. Identification of species is impossible without expert help, and often requires examination of a sample in a lab.

Brown tube sponge
Tubes or tall pipes are another common form for sponges, with very large specimens such as the organ-pipe sponge reaching almost 2 meters in height. Water passes through the sponge wall, and out through the top.

Hanging vine algae

Y branched algae

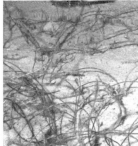
Manatee grass

Turtlegrass

Red mangrove

Black mangrove

Lavender rope sponge

Brown tube sponge

Star encrusting sponge *Halisarca* sp

Many sponges encrust the sea floor (and some will even smother and kill corals). On this image it is possible to see the fine pores, into which water is drawn, and the larger pores where the water flows out, having been filtered for food.

Azure vase sponge *Callyspongia plicifera*

Perhaps the most beautifully colored sponge on the reef – an iridescent pinky blue. Like many sponges this forms a vase or goblet shape, its outer surface finely sculpted with convoluted ridges and valleys.

CORALS AND THEIR RELATIVES: CNIDARIA

Giant anemone *Condylactis gigantea*

The sea anemones are giant solitary polyps, related to the corals, and the giant anemone is the largest of all in the Caribbean, reaching more than 30 centimeters across. The large tentacles (sometimes with swollen pink tips) are packed with stinging nematocysts.

Bent sea rod *Plexaura flexuosa*

Several soft corals are commonly called sea rods – these are branching, with the branch tips growing vertically upwards. The bent sea rod is more highly branching than most, and the branches lie in a single plane, like a fan.

Deepwater sea fan *Icilgorgia schrammi*

Although related to the sea rods and plumes, the sea fans have considerably more stiffening in their skeletons. The deepwater sea fan thrives on steep slopes and walls, often below 20 meters, and in areas where currents sometimes run.

Common sea fan *Gorgonia ventalina*

Abundant on shallow reef slopes, these pretty corals may be purple, yellow, or pale brown. They thrive in areas of gentle water movements.

Fire coral *Millepora alcicornis*

The fire corals are quite a distinct group of corals. They have limestone skeletons, but are in a separate group from the stony corals. They have powerful stinging organs which can produce a significant rash if they touch human skin. The branching fire coral thrives in shallow reef areas, even in quite rough water.

Bushy black coral *Antipathes* sp

Black corals come in a range of forms from the single-strand wire coral to feather, bushy, and fan forms. They lay down a dense skeleton of protein which is very durable and has been important in the jewelry trade. Looking up close, each polyp has six permanently extended tentacles. They do not have algae in their tissues and are often found in deeper water.

Star encrusting sponge

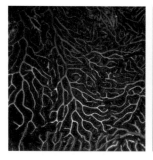

Azure vase sponge

Giant anemone

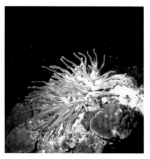

Bent sea rod

Deepwater sea fan

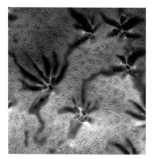

Common sea fan

Fire coral

Bushy black coral

Sponge zoanthid *Parazoanthus parasiticus*

Zoanthids are like very small, simple anemones with just a single ring of tentacles. They are usually found in colonies. Some are densely packed, with short tentacles, and survive in the most wave-swept areas of the reef. This sponge zoanthid lives on the surface of sponges – it could be parasitic, but some have suggested that it may protect the sponges from certain predators.

Stony corals

Elkhorn coral *Acropora palmata*

Generally found in shallow water and quite resistant to wave action. Together with the staghorn coral this branching coral species is one of the most important corals in building reefs. Numbers have collapsed in recent years as a result of white-band disease, but in a few areas this spectacular coral is making a good recovery.

Great star coral *Montastrea cavernosa*

There are few more distinctive corals than the great star coral, with its massive form and clearly defined bumps formed by the individual polyps. Like most stony corals it keeps its polyps withdrawn during the day and extends them only at night.

Smooth flower coral *Eusmilia fastigiata*

Not a common coral, but often seen in protected waters below about 7 meters. The individual polyps are quite large and widely separated, each building up a long tubular skeleton.

Pillar coral *Dendrogyra cylindrus*

A beautiful coral, usually found on flat or gently sloping reef, and often surrounded by soft corals. Colonies can reach 3 meters in height. The polyps usually extend their tentacles during the day.

Sheet coral *Agaricia lamarcki*

This is one of several closely related corals. It grows either by encrusting the sea floor or extending in wide circular sheets, usually on steeper slopes and walls. The closely related lettuce corals favor bright sunlit waters and grow well in wave-swept environments.

Symmetrical brain coral *Diplora strigosa*

A number of massive or mound-shaped corals are called brain corals and have surfaces covered in fine grooves forming wonderful reticulated patterns (very similar to the grooves on the surface of a brain). Close inspection reveals the small holes of individual coral polyps lying within the grooves.

Finger coral *Porites porites*

This is a fairly widespread branching coral found on most zones of the reef. The small polyps are embedded within the skeleton, giving a fairly smooth surface, although quite often they extend their tentacles during the day.

Sponge zoanthid

Elkhorn coral

Great star coral

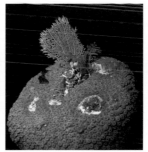

Smooth flower coral

Pillar coral

Sheet coral

Symmetrical brain coral

Finger coral

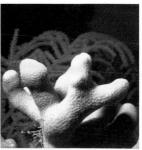

Maze coral *Meandrina meandrites*

An attractive coral which may encrust the sea floor or grow into plate-like or even domed formations.

SEGMENTED WORMS: POLYCHAETA

Split crown feather duster *Anamobaea orstedii*

There are several very similar solitary feather duster worms. All extend their exquisite mouthparts from a small tube, often hidden in coral and rocks. Sensitive to water movements and shadows, they retreat instantaneously if approached too close.

Social feather duster *Bispira brunnea*

This worm always grows in clusters and the thin tubes which the worms build around their bodies are clearly visible. The color of their feeding fans varies considerably from place to place across the Caribbean.

CRUSTACEANS: CRUSTACEA

Spiny lobster *Panulirus argus*

Very familiar on reefs and restaurant tables, spiny lobsters hide in recesses through the day and are very wary of humans, but emerge at night to range across the reef feeding on small invertebrates.

Giant hermit crab *Petrochirus diogenes*

Hermit crabs are an unusual group of crabs with relatively soft and elongated bodies which they protect in empty snail shells. They are scavengers, feeding on dead organic matter, and as they grow they have to find larger shells. The giant hermit crab often moves into conch shells on areas of sand and seagrass.

Pederson cleaner shrimp *Periclimenes pedersoni*

These delicate, translucent shrimps are to be found sheltering in the tentacles of sea anemones, often waving their bodies and swaying their antennae to attract fish to their cleaning services. This is a female with pink eggs attached to her belly.

Fiddler crab *Uca sp*

Walking through areas of mangroves at low tide you may see the remarkable form of the fiddler crab. Males of the species are equipped with one greatly elongated pincer which they use both to attract the attention of females and to warn other males to stay out of their territory.

King crab *Mithrax spinosissimus*

King crabs go under a host of other names, including spider crab and channel clinging crab. They are giants, their carapaces measuring up to 17 centimeters across, and reaching almost a meter in diameter with their arms outstretched. They leave their dark recesses at night to go in search of algae, their main food.

Maze coral

Split crown feather duster

Social feather duster

Spiny lobster

Giant hermit crab

Pederson cleaner shrimp

Fiddler crab

King crab

MOLLUSKS: MOLLUSCA

Queen conch *Strombus gigas*
Probably the best known, and undoubtedly the most commercially important, gastropod on the reef. Queen conch are heavily armored and slow moving grazing snails. They have a pair of eyes on stalks to keep watch and withdraw into their shells when approached.

Flame helmet *Cassis flammea*
Helmet shells are chunky but very beautiful – the flame helmet typically reaches 11 centimeters in length, but the similar emperor helmet has been recorded to 25 centimeters. In stark contrast to conch, these are voracious predators, and regularly feed on sea urchins which they are able to overpower with surprising agility, undaunted by the urchin's sharp spines.

Florida regal sea goddess
Hypselodoris edenticulata
The beautiful sea goddesses are part of a group of shell-less snails known as nudibranchs, which means "naked gills" – the rosette or flower on their backs is indeed a set of gills. They feed on sponges and are able to store up the toxins from their prey to use against any creatures which try to eat them.

Flame scallop *Lima scabra*
Also known as the rough fileclam, these beautiful bivalves are often overlooked as they nestle in crevices on the reef. They are filter feeders and their bright red coloration comes from the hemoglobin in their blood. They tend to clamp tightly shut if approached too close.

Caribbean reef squid *Sepioteuthis sepioidea*
Like their relations the octopus, squid are highly evolved mollusks. Squid are fast, active hunters. They normally move with fine undulations of the fins along their sides, but if threatened can put on great bursts of speed by squirting water from their body cavity in a form of jet propulsion.

ECHINODERMS: ECHINODERMATA

Sponge brittle star *Ophiothrix suensonii*
Although similar to starfish, brittle stars have long thin legs radiating out from a central disc. This particular species is a filter feeder, and always associated with sponges, perhaps using water flows generated by the sponge to intercept food particles from the plankton.

Giant basket star *Astrophyton muricatum*
Basket stars are closely related to the brittle stars, but their very slender arms are finely branched. Hidden within the reef's recesses during the day, they clamber out to prominent points at night and unfurl their fabulous, highly mobile arms to filter the water for food.

Queen conch

Flame helmet

Florida regal sea goddess

Flame scallop

Caribbean reef squid

Sponge brittle star

Giant basket star

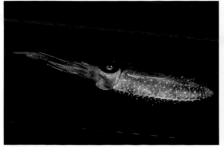

Cushion sea star *Oreaster reticulatus*

A large starfish with thick, short arms, reaching 40 centimeters across. Found over sandy areas and in seagrass beds, they move slowly about looking for food. They mainly feed on bivalves which they are able to hold open with their powerful muscles. They then evert their stomachs into the shells and digest them outside of their own bodies.

Reef urchin *Echinometra viridis*

Sea urchins have a highly developed skeleton made of almost solid calcium carbonate plates, and they are further defended by long spines. They are important algal grazers – they have a strange feeding apparatus called an Aristotle's lantern on their underside, with five powerful and continuously growing teeth. The reef urchin, like most echinoderms, is nocturnal, and only seen hidden in crevices during the day.

Furry sea cucumber *Astichopus multifidus*

Sea cucumbers are large sausage-shaped echinoderms with a mouth at one end and an anus at the other. Most, such as the furry sea cucumber, move laboriously over sandy areas feeding on detritus.

Tiger tail cucumber *Holothuria thomasi*

A strange and surprisingly dynamic sea cucumber, this species can reach 2 meters in length. They tend to keep their rear end fixed in the reef while the front end moves over the substrate like a vacuum cleaner, looking for food.

VERTEBRATES: VERTEBRATA

Fishes

Caribbean reef shark *Carcharhinus perezi*

One of the more widespread sharks, particularly around islands. A fairly heavy-bodied species, typically reaching 2-3 meters maximum length. Generally wary of people, but has been known to be aggressive, especially if there are spearfishers in the water.

Silky shark *Carcharhinus falciformis*

An oceanic species, but known to come to reefs where these are close to deep water. This is a large shark, but slender, and is known to reach over 3 meters in length.

Nurse shark *Ginglymostoma cirratum*

One of the most common and unmistakable reef sharks, with brown skin and a heavy body. They have been recorded to 4 meters in length. Most often seen resting on the bottom, sometimes with their head under overhangs. They eat lobsters and other invertebrates.

Cushion sea star

Reef urchin

Furry sea cucumber

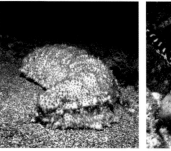

Tiger tail cucumber

Caribbean reef shark

Silky shark

Nurse shark

Yellow stingray *Urolophus jamaicensis*

A small stingray, reaching to 35 centimeters across. Quite variable in color, but marked with an array of pale and dark spots or reticulations. The double spine on its tail can inflict a nasty wound, but is only used in defense. Feeds on shrimps and other small invertebrates.

Green moray *Gymnothorax funebris*

Green morays are the largest of the Caribbean morays, reaching to 180 centimeters. They are active predators, working by day or night, snaking their way through the reef and using a highly acute sense of smell to locate fish and crustaceans.

Tarpon *Megalops atlanticus*

A giant inshore fish, reaching 2.4 meters, regarded as an ancient and primitive form of fish. Often found in very shallow water, even venturing into estuaries and rivers, it also swims out over reefs, sometimes in large schools. Feeds on smaller sardine-like fish.

Longspine squirrelfish *Holocentrus rufus*

Large eyes and red coloration are clear indicators of a nocturnal fish – large eyes for improved vision and red because in the filtered light of the sea this blends in with the shadows. Squirrelfish are often seen during the day lurking in dark areas, and feed from dusk to dawn on small invertebrates.

Glasseye snapper *Heteropriacanthus cruentatus*

Another nocturnal species, belonging to a family known as bigeyes. Usually an even red color, they can quickly adopt this blotched pattern. They are active predators of invertebrates and small fish.

Trumpetfish *Aulostomus maculatus*

The trumpetfish is a predator of fish and invertebrates found on almost all Caribbean reefs. It uses stealth to get close to its prey – an adept color changer, it also often hides, vertically, amongst soft corals, or "shadows" other fish, and is capable of opening its narrow mouth surprisingly wide.

Spotted scorpionfish *Scorpaena plumeri*

Masters of camouflage, scorpionfish are often overlooked by divers and snorkelers. They are classic sit-and-wait predators, and when an unsuspecting fish comes past will quickly lunge out to catch it. They get their name from the venomous spines on their dorsal fins.

Nassau grouper *Epinephelus striatus*

The groupers are part of the seabass family and are large predators with heavy jaws. The Nassau grouper is one of the best known and most beautiful. Its numbers are much diminished in many areas due to overfishing. Reaches 1 meter.

Yellow stingray

Green moray

Tarpon

Longspine squirrelfish

Glasseye snapper

Trumpetfish

Spotted scorpionfish

Nassau grouper

Jewfish *Epinephelus itajara*
While sharks are the largest visitors, the jewfish is the largest fish to make its home on a reef – with the largest specimen recorded to 2.5 meters and a weight of 455 kilos. They mostly feed on fish, but have been found with young turtles and stingrays in their stomachs. Often inquisitive, they have suffered from fishing and spearfishing.

Red hind *Epinephelus guttatus*
Medium-sized grouper, common in areas where fishing pressure is not great. The reddish spots and dark edges to the rear fins and tail are distinctive. This is often a fairly inquisitive fish and will sometimes come up to inspect divers. Reaches over 60 centimeters.

Tiger grouper *Mycteroperca tigris*
Relatively common on many reefs. Like all in the grouper family, it can change its background colors considerably, but the diagonal, slightly paler, lines across its back are almost always visible. Reaches 1 meter.

Yellowfin grouper *Mycteroperca venenosa*
A large grouper, its sides are marked with big oval blotches on a slightly lighter background, but these fish can darken to nearly black, or become very reddish in color. The faintly yellow edge to the pectoral fin is almost always visible. Reaches over 90 centimeters.

Coney *Cephalopholis fulvus*
A small, slender grouper which is abundant on most reefs, and still found even on overfished reefs. They have an enormous range of colors, from bright yellow, to a bicolor phase with a red upper half and white lower half, to a red color all over. All are usually marked with very small electric blue spots ringed in black. Reaches 40 centimeters.

Yellowbelly hamlet *Hypoplectrus aberrans*
The diminutive, but beautiful, hamlets are in the same family (seabasses) as the groupers. Although only reaching 10-15 centimeters they are still active predators, and have the same protruding lower jaw as their larger relatives.

Fairy basslet *Gramma loreto*
Also known as the royal gramma, this is one of the gems of the coral reef, with its unmistakable violet and yellow colors. They live in colonies of ten or more fish and drift out from the reef to feed on plankton. They are also often seen in caves and overhanging areas where they swim with their bellies against the surface of the rock, which means swimming upside-down on the cave roof.

Jewfish

Red hind

Tiger grouper

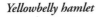

Yellowfin grouper   *Coney*

Yellowbelly hamlet

Fairy basslet

Horse-eye jack *Caranx latus*

The jacks are powerful predators that rely on speed, not stealth, to catch their prey. Horse-eye jacks are wide ranging and often sweep in to a reef from offshore waters. They travel and hunt in packs. Reaches 80 centimeters.

Bar jack *Caranx ruber*

The commonest of the reef jacks, bar jacks are often seen alone or in small groups, and are marked by their distinctive electric blue backs fringed with a black streak that drops across to the lower half of the tail. These colors tend to fade a little in larger fish. Reaches 40 centimeters.

Mutton snapper *Lutjanus analis*

The snappers are a group of moderate to large fish, generally pale or silvery in color with long pointed snouts. They are all active predators of invertebrates and small fish. The mutton snapper has a small black spot on its flank and faint blue markings below its eye. Reaches 80 centimeters.

Schoolmaster *Lutjanus apodus*

Undoubtedly the most abundant of the Caribbean snappers, schoolmasters are often found in large schools, sometimes mixed with grunts. Younger fish are commonly seen among mangroves. Reaches 65 centimeters.

Cubera snapper *Lutjanus cyanopterus*

The largest of the snappers, recorded to 1.5 meters and 57 kilos in weight. They may be solitary or found in small groups, but like many of the larger snappers they travel long distances to form very large spawning aggregations once or twice a year.

Bluestriped grunt *Haemulon sciurus*

The grunts are similar in profile to the snappers, but with slightly less elongated snouts. They get their name because they make a grunting sound by grinding their teeth. Bluestriped grunts are distinctive, with a black tail and fine blue stripes, and often form large schools, sometimes mixed with other grunt species.

French grunt *Haemulon flavolineatum*

A smaller grunt, typically reaching 25 centimeters. Like other grunts they mostly rest through the day in large schools, but become active at dusk when they head out to feed on a variety of invertebrates.

Horse-eye jack *Bar jack* *Mutton snapper* *Schoolmaster*

Cubera snapper *Bluestriped grunt* *French grunt*

White margate *Haemulon album*

A large grunt with a distinctive high back and white, almost silvery color, but a dusky tail. Typically found alone or in small groups. They are known to eat crustaceans, small fish, and even sea cucumbers. Reaches 70 centimeters.

Porkfish *Anisotremus virginicus*

A common and quite unmistakable fish with its bright white and yellow striped body and twin black bars on its head. These grunts are usually solitary, but often mix in with schools of other grunt species. The young often act as cleaner fish, picking parasites from other species.

Pluma *Calamus pennatula*

This is one of the porgies (a group of silvery fish with quite highly arched backs). It is widespread in the Bahamas and eastern Caribbean, where it feeds on invertebrates and occasionally digs in the sand for food. Reaches 35 centimeters.

Saucereye porgy *Calamus calamus*

These large-eyed porgies, like many of their relatives, are highly curious. If divers keep still they will approach to see what is happening. They feed on a varied diet of worms, mollusks, even sea urchins. Reaches 40 centimeters.

Spotted drum *Equetus punctatus*

The spotted drum is one of the brightest members of a large family – the adults are smartly marked with black and white stripes, and with polka-dot spots on their rear fins and tail, but the juveniles, which are more commonly observed, are simply striped, with a wonderful high-plumed dorsal fin and an elongated tail. They rest out the day in recesses, but allow divers to come quite close.

Yellow goatfish *Mulloidichthys martinicus*

Goatfishes in general are elongate fish which get their name from the long and prehensile barbels under their chin, said to resemble a goat's beard. These barbels are highly sensitive and can be flicked down to probe the sand, where they quickly detect small invertebrates to eat.

Banded butterflyfish *Chaetodon striatus*

Butterflyfish are a group of small, disc-shaped fish found only on coral reefs. The banded butterflyfish is unmistakable, moving about the reef in pairs, often in shallow water.

White margate

Porkfish

Pluma

Saucereye porgy

Spotted drum

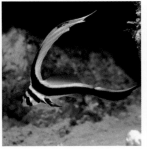

Yellow goatfish

Banded butterflyfish

Spotfin butterflyfish *Chaetodon ocellatus*
The largest of the Caribbean butterflyfish, reaching to 20 centimeters. All butterflyfish have fine mouthparts which they use to pick at small invertebrates and the individual polyps of corals, but each species has favored food items. Spotfin butterflyfish often pick at tubeworms and even take small shrimps.

Queen angelfish *Holocanthus ciliaris*
Angelfish are closely related to butterflyfish, and similarly brightly colored. They feed almost entirely on sponges. The queen angelfish is regarded as one of the most beautiful fish on the reef, and is often found skulking in recesses.

Rock beauty *Holocanthus tricolor*
A highly distinctive angelfish, the rock beauty is often seen alone. Large males defend territories which include the territories of several smaller females, a dispersed harem.

French angelfish *Pomacanthus paru*
A widespread species, French angelfish are often quite unafraid of divers. Like many other angelfish and butterflyfish they remain together as mating pairs, for extended periods of months or even years.

Gray angelfish *Pomacanthus arcuatus*
Very closely related to the French angelfish, gray angelfish have similar habits. Both species feed primarily on sponges, but also add soft coral polyps and quite a number of other invertebrates to their diets. Reaches 36 centimeters.

Sergeant major *Abudefduf saxatilis*
The damselfish are the most abundant fish on many coral reefs, and their presence is often taken for granted. Hard to miss, however, is the smartly striped sergeant major, a plankton feeder often found in large groups high above the reef and totally unafraid of divers and snorkelers.

Blue chromis *Chromis cyanea*
Another striking damselfish, the blue chromis is often seen in large numbers swimming above the reef, picking at plankton. Being very small they pick not only at zooplankton, but also tiny pieces of phytoplankton. A close relative, the brown chromis, is probably the most abundant species in the Caribbean.

Spotfin butterflyfish

Queen angelfish

Rock beauty

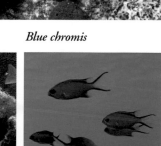

French angelfish

Gray angelfish

Sergeant major

Blue chromis

Spanish hogfish *Bodianus rufus*

The wrasse family is a very diverse group of fish. The purple and golden-yellow Spanish hogfish is one of the most attractive. The juveniles act as cleaner fishes, picking parasites off other fish. At about 15 centimeters in length the females, like all wrasses, undergo a sex change, and so all larger specimens are males.

Puddingwife *Halichoeres radiatus*

The wrasses of this genus are all elongate, cigar-shaped fish, but the puddingwife is much larger than its relatives, reaching 50 centimeters. Like some other large wrasses it feeds on mollusks, sea urchins, and crustaceans.

Bluehead wrasse *Thalassoma bifasciatum*

Found on almost every coral reef in the Caribbean, the bluehead wrasse is a well known, and well studied character. The males, with their flag-like markings, patrol the reef, sometimes alone, but often joining small gangs of the smaller, bright yellow females. They feed on invertebrates and regularly raid the nests of fish which lay eggs on the sea floor.

Hogfish *Lachnolaimus maximus*

Male hogfish have been recorded to 91 centimeters, and are the only wrasses with these spectacular elongate spines on their dorsal fin. The photo below was taken at night and the fish has adopted a mottled color to blend in with its surroundings.

Stoplight parrotfish *Sparisoma viride*

The parrotfish are close relatives of the wrasses, but marked out by the adaptation of their teeth into a powerful beak for scraping algae off the rocks. Mature male stoplight parrotfish always show yellow on their tails and a small yellow spot above their gill cover.

Princess parrotfish *Scarus vetula*

Almost all parrotfish undergo remarkable transformations of color as they mature. Mature males are typically the brightest, often blueish or greenish. The princess parrotfish has a distinctive yellow streak along its side.

Spanish hogfish

Puddingwife

Bluehead wrasse

Hogfish

Stoplight parrotfish (mature male)

Princess parrotfish (mature male)

Queen parrotfish *Scarus vetula*

Immature male and female parrotfish are typically not as bright as the mature males, but usually have their own distinctive patterns.

Immature parrotfish *Scarus* sp

Young parrotfish can be difficult to distinguish. Several species have fine white and dark stripes, and the different species sometimes swim together in small schools.

Barracuda *Sphyraena barracuda*

The great barracuda is an awe-inspiring fish, with its sleek, silvery body, low underslung jaw, and its sharp peg-like teeth. It is a powerful predator of other fish, with a rather disconcerting habit of approaching divers and snorkelers and simply watching them. There are, however, no records of unprovoked attacks on humans. Reaches 2 meters.

Broadstripe goby *Gobiosoma prochilos*

This is one of several cleaner gobies. All are small (reaching 4 centimeters), and many are similar looking. They spend the day resting in small groups in prominent locations waiting for custom. Much larger fish come to these cleaning stations and the gobies spring into action, working their way over the fish picking off parasites, dead skin and the like, which they eat.

Roughead blenny *Acanthemblemaria aspera*

The blennies are another group of tiny fish, characterized by large blunt heads and large eyes. Many remain virtually hidden in small holes, and are often overlooked. They feed on tiny crustaceans.

Blue tang *Acanthurus coeruleus*

The blue tang is part of the surgeonfish family, so-named because of a sharp blade like a scalpel, which lies in a small groove near the base of the tail (yellow in this species). By twisting their tails the blade comes out and is used as a form of defense. They are algal grazers and roam the reef in large schools.

Doctorfish *Acanthurus chirurgus*

Doctorfish are another surgeonfish, again often seen in schools. They change color frequently, and the ten fine lines on their sides can become almost invisible.

Queen parrotfish (female / immature male)   *Parrotfish (immature)*   *Barracuda*

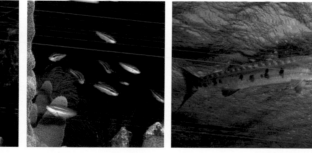
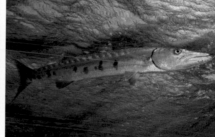

Broadstripe goby   *Roughead blenny*   *Blue tang*   *Doctorfish*

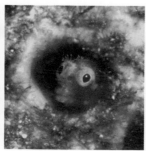

Whitespotted filefish *Cantherhines macrocerus*

The filefish, and a similar group, the triggerfish, have a strong spine on their backs which is usually kept folded down, but can be lifted and locked upright. When threatened they retreat into a narrow space and use this spine to lock themselves in so that predators cannot pull them out. The whitespotted filefish is a master color changer and can turn its spots on or off in a matter of seconds.

Smooth trunkfish *Lactophrys triqueter*

The scales of the smooth trunkfish and its relations have been modified into a bony armor and only the eyes, mouth, fins, and tail are free to move. They swim slowly looking for small invertebrates on the sea floor.

Balloonfish *Diodon holocanthus*

There are four species of porcupinefish commonly seen in the Caribbean. All have a body covering of sharp spines and will, if threatened, inflate their bodies to make them look larger and harder to swallow with spines pointing outwards. The balloonfish is distinguished by having quite long spines, which even extend on to its head.

Marine turtles

Five marine turtles are found across the Caribbean. Kemp's ridley turtles are the smallest and rarest, with a heart-shaped gray to green shell. Green turtles have quite rounded heads, and pale undersides. They are not green, but brownish, and get their names from the color of their flesh. Loggerhead turtles are large, and have a very big, heavy-looking head. A sixth species, the olive ridley, is occasionally seen in the southern Caribbean.

Hawksbill turtle *Eretmochelys imbricata*

One of the most commonly encountered turtles. They are so-called because the "beak" around the mouth is often hooked like a bird of prey. Their shells are beautifully marked in streaks of brown and orange, although this can be obscured, particularly in larger individuals. The rear of the shell has a serrated edge.

Leatherback turtle *Dermochelys coriacea*

The rarest of the Caribbean turtles and highly threatened worldwide, these wonderful giants have shells reaching to nearly 2 meters long, with ridges running along their length. Rarely seen on reefs, these are open ocean turtles, known to migrate for thousands of kilometers and feeding on jellyfish and other larger creatures in the plankton.

Whitespotted filefish, with spots turned "on" and "off"   *Smooth trunkfish*   *Balloonfish*

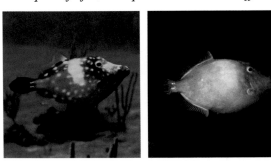

Hawksbill turtle   *Leatherback turtle*

SELECTED ENVIRONMENTAL ORGANIZATIONS

NORTHERN CARIBBEAN
BERMUDA
Bermuda Audubon Society
www.audubon.bm
PO Box HM 1328, Hamilton HM FX
Promotes conservation through education
and outreach. Also runs a series of
terrestrial and coastal reserves.

Bermuda Biological Station for Research
www.bbsr.edu/
Ferry Reach, St George's, GE 01
An important research laboratory, also
engaged in education projects at school
and university level.

Bermuda National Trust
www.bnt.bm
PO Box HM 61, Hamilton, HM AX
Natural and cultural conservation work,
including the management of nature
reserves and some coastline areas, and
education programs.

Bermuda Zoological Society
www.bamz.org/
PO Box FL 145, Flatts, FL BX
Works with the Department of
Conservation Sciences and the Bermuda
Aquarium Museum and Zoo in education,
research and conservation work.

BAHAMAS
Andros Conservancy and Trust (ANCAT)
www.ancat.org/
Fresh Creek, Andros Island
Works exclusively on land and sea areas in
Andros and with the Bahamas National
Trust in management of protected areas.

Bahamas National Trust
www.bahamas.gov.bs/bahamasweb/
visitingthebahamas.nsf/subjects/national+
trust
PO Box N-4105, Nassau
Established with government support,
manages the terrestrial and marine
protected areas across the Bahamas.
Actively supporting species recovery
programs, education, and some
research.

Bahamas Reef Environment Educational Foundation (BREEF)
www.breef.org
PO Box N-7776, Nassau
Promotes conservation through education,
research, and management. Is building
support for the establishment of no-take
marine reserves throughout the Bahamas.

USA
Florida Keys National Marine Sanctuary
http://floridakeys.noaa.gov/
PO Box 500368, Marathon FL 33050
The largest coral reef protected area in
Florida, the park management body
oversees regulations, undertakes research,
and provides important educational and
outreach resources.

Florida Marine Research Institute
www.floridamarine.org/
*100 Eighth Avenue SE, St Petersburg
FL 33701-5095*
A state supported research institution with
important research and information
resources, conducting long-term
monitoring all along the Florida Reef Tract.

Rosenstiel School of Marine and Atmospheric Science (RSMAS)
www.rsmas.miami.edu
*University of Miami – RSMAS, 4600
Rickenbacker Causeway, Miami, FL 33149*
The University of Miami's main teaching
and research school, and a leader in
oceanographic research. Includes the
National Center for Caribbean Coral Reef
Research (NCORE – www.ncoremiami.
org).

US National Aeronautics and Space Administration (NASA)
http://eol.jsc.nasa.gov
*Earth Sciences and Image Analysis, NASA-
Johnson Space Center, Houston, Texas*
NASA provides considerable information
on coral reefs, including, through its Earth
Sciences and Image Analysis Laboratory,
large numbers of images taken by
astronauts and used both by scientists and
for educative purposes.

US National Oceanographic and Atmospheric Administration (NOAA)
www.noaa.gov/ocean.html and
www.coralreef.noaa.gov/
*1305 East-West Highway, Silver Spring
MD 20910*
The US government's primary agency
dealing with atmospheric and
oceanographic issues nationally and
internationally. Has taken a leading role in
coral reef conservation both nationally (it
runs the National Marine Sanctuaries) and
internationally, with ongoing work on
marine protected areas, coral diseases and
bleaching, reef mapping, and much more.
Offers important information resources
and networking, especially through its
internet facilities.

TURKS AND CAICOS
School for Field Studies – Center for Marine Resource Studies
www.fieldstudies.org
*CMRS, South Caicos, 1 West St, South
Caicos*
An international research and educational
lab based on South Caicos, mostly
catering to university students from North
America and conducting various research
projects on the reefs and adjacent
ecosystems.

Turks and Caicos National Trust
www.turksandcaicos.tc/NationalTrust/
PO Box 540, Providenciales
Aims to conserve the natural and cultural
heritage, working in land and coastal
protected areas and with education and
community outreach projects.

MEXICO AND CENTRAL AMERICA
MEXICO
Amigos de Sian Ka'an
www.amigosdesiankaan.org
*Crepúsculo No 18 esq. Amanecer, SM 44,
Mza 13, Fracc. Alborada, Cancún,
Quintana Roo, 77506*
Works to conserve the natural heritage of
the state of Quintana Roo – identifying
areas for protection, preparing
management plans, carrying out field
surveys (including coral reefs), publishing a

variety of materials, and training local communities in subjects from fisheries to tourism guiding.

Centro de Investigación y de Estudios Avanzados del Instituto Politécnico Nacional (Cinvestav)

www.mda.cinvestav.mx
Cinvestav Unidad Mérida, Carretera Antigua a Progreso Km. 6, Apartado Postal 73 Cordemex, Mérida, Yucatán, CP 97310
Research center with a department of marine resources conducting research and education, particularly in the waters around the Yucatán.

Centro Ecológico Akumal

http://ceakumal.org/
Apartado Postal 2, Akumal, Quintana Roo, 77760
Facilitates academic research, environmental education, cultural exchange, and influencing policy.

Centro Mexicano de Derecho Ambiental (CEMDA)

www.cemda.org.mx/
Atlixo, Col. Condesa, 6148

Instituto de Ciencias del Mar y Limnología, Universidad Nacional Autónoma de México, Unidad Académica de Sistemas Arrecifales

www.icmyl.unam.mx/arrecifes/rsu.html
Ap. Postal 1152, CP 77500 Cancún, Quintana Roo
Runs the research station near Puerto Morelos, as well as research and education on all Mexico's Caribbean reefs.

BELIZE
Belize Audubon Society

www.belizeaudubon.org/
12 Fort St, PO Box 1001, Belize City
Manages eight protected areas including Blue Hole National Monument and Half Moon Caye. Actively involved in education projects, research, and advocacy.

Coastal Zone Management Authority and Institute

www.coastalzonebelize.org/
PO Box 1884, Princess Margaret Drive, Belize City
Government body working in the coastal zone, particularly to conduct marine research, maintain a data center, organize training courses, support other agencies involved in coastal zone management (CZM), and assist with preparation of a national CZM plan.

HONDURAS
Bay Islands Conservation Association

Actively promotes marine conservation through the Bay islands. Led in the development of the Sandy Bay West End Marine Reserve.

COSTA RICA
Centro de Investigación en Ciencias del Mar y Limnología (CIMAR), Universidad de Costa Rica

www.cimar.ucr.ac.cr
Ciudad de la Investigación, Universidad de Costa Rica, San Pedro, San José, 2060
Principal marine research center in the country.

PRETOMA

www.tortugamarina.org
PRETOMA, 1203-1100, Tibás, San José
Actively campaigns for marine turtle protection and the marine environment more generally. Working with communities and volunteers, and taking legal action over critical issues.

Promar Foundation (PROMAR)

www.promar.or.cr
PO Box 11709-1000, San José
Focuses on education and on cetaceans.

PANAMA
Asociación Nacional para la Conservación de la Naturaleza (ANCON)

www.ancon.org/
Apartado Postal 1387, Panamá 1
Working on biodiversity conservation through education, advocacy, and research.

Smithsonian Tropical Research Institute

www.stri.org
Smithsonian Tropical Research Institute, Roosevelt Avenue, Ancon, Balboa – Tupper Building, PO BOX 2072
Research organization linked to the US-based Smithsonian Institution, operates field stations, among others, on Bocas del Toro and Isla Galeta.

GREATER ANTILLES
CUBA
Instituto de Oceanología

www.cuba.cu/ciencia/citma/ama/oceanologia/Default.html
Avenue 1ra No. 18406 entre 184 y 186, Reparto Flores, Playa, La Havana
The key marine research institution linked to the Ministerio de Ciencia, Tecnología y Medio Ambiente. Undertakes research into biodiversity, fisheries and sustainable use of natural resources.

CAYMAN ISLANDS
Cayman Islands Department of Environment

www.doe.8m.com/doewebsite/doe.html
PO Box 486GT, Grand Cayman
Government department, running the marine parks, undertaking reef research, a Marine Turtle Programme, and some excellent educational projects.

Cayman Turtle Farm

www.turtle.ky/
PO Box 645, George Town, Grand Cayman
Probably the only successful captive breeding project for turtles in the world. Releases considerable numbers into the wild each year.

The National Trust for the Cayman Islands

www.nationaltrust.org.ky/
PO Box 31116SMB, Grand Cayman
Aims to preserve natural and cultural heritage, runs land and wetland protected areas, and education programs. Supports some research.

JAMAICA
Caribbean Coastal Area Management (C-CAM) Foundation

www.portlandbight.com.jm/
PO Box 33, Lionel Town, Clarendon
Responsible for managing the Portland Bight Protected Area, which has led the way in getting local people, especially fishers, involved in marine protection.

Centre for Marine Sciences – University of the West Indies

www.uwimona.edu.jm/cms/
A research and educational establishment; also runs the Discovery

Bay Marine Lab and the Caribbean Coastal Data Centre.

Montego Bay Marine Park Trust
www.mbmp.org/
Pier 1, Howard Cooke Blvd, Montego Bay
Aiming to conserve, restore, and manage this important park, and to involve the local community and traditional users.

Negril Coral Reef Preservation Society
PO Box 2725, Negril, Westmoreland
Has played a key role in the establishment and management of the Negril Marine Park and continues to work with community education, advice to developers, and coral reef monitoring.

HAITI
Foundation for the Protection of Marine Biodiversity (Fondation pour la Protection de la Biodiversité Marine) (FoProBiM)
www.foprobim.org/
PO Box 642, Port-au-Prince, Haiti or 7812 Green Twig Road, Bethesda MD 20817-6918, USA
Seeking to instigate marine conservation at the community level. Undertakes field research, organizes beach clean-ups.

DOMINICAN REPUBLIC
Center for the Conservation of Samaná Bay and its Surroundings
www.samana.org.do/
Av. La Marina, Tiro al Blanco, Centro Para La Naturaleza, Apdo 243, Samaná

PUERTO RICO
CORALations
www.coralations.org
CORALations, Inc., PO Box 750, Culebra, PR 00775
Working to conserve and restore reefs, particularly through local communities and through education projects.

Sociedad Ambiente Marino
http://web.uprr.pr/sam/
PO Box 22158 SJ, PR 00931-2158
Activities include: coastal clean-ups, reef monitoring, training, and working with the government to encourage better legal and management systems for the environment.

US VIRGIN ISLANDS
Friends of the Virgin Islands National Park
www.friendsvinp.org
PO Box 811, St John, USVI 00831
Coordinates fund-raising, supports outreach and education programs and the work of volunteers in the park.

University of the Virgin Islands – Center for Marine and Environmental Studies
http://marsci.uvi.edu/
2 John Brewer's Bay, St Thomas, 00802
Coordinates marine research in the university, as well as outreach programs, with research bases on three main islands.

BRITISH VIRGIN ISLANDS
Association of Reef Keepers
www.arkbvi.org
PO Box 3169, PMB 130, Road Town, Tortola
Has produced important promotional materials on reef protection, conducts basic research, and is working with a subsidiary, Island Erosion, to educate developers on the problems of sedimentation.

British Virgin Islands National Parks Trust
www.bvinationalparkstrust.org/
#61 Main St, Road Town, Tortola
The body which manages national parks in the country. Has worked on species restoration projects and the installation of mooring buoys.

LESSER ANTILLES
ANGUILLA
Anguilla National Trust
http://web.ai/ant/
PO Box 1234, The Valley
Involved in natural and cultural conservation work. Runs the protected areas in the country. Involved in environmental education and turtle monitoring.

NETHERLANDS ANTILLES (NORTHERN GROUP)
Nature Foundation of Sint Maarten (NAFSXM)
www.naturefoundationsxm.org
PO Box 863, Philipsburg, St Maarten
Working to protect the natural environment through the proposed marine park, and a terrestrial park. Involved in education, outreach to communities, and advocacy.

Netherlands Antilles Coral Reef Initiative
www.nacri.org/
A collaboration of environmental, government and industrial groups with an interest in coral reefs.

Saba Conservation Foundation
www.sabapark.org
PO Box 18, The Bottom, Saba
Manages the marine park and national parks; promotes research and education.

St Eustatius National Parks Foundation (STENAPA)
www.statiapark.org
St Eustatius National and Marine Park Office, Gallows Bay, St Eustatius
Manages the marine park and national parks, and undertakes reef research. Also works with volunteers, and is actively engaged in education on the island.

ST KITTS AND NEVIS
Nevis Historical and Conservation Society
www.nevis-nhcs.org/
PO Box 563, Charlestown, Nevis
A society interested in historical, cultural, and natural conservation.

St Christopher Heritage Society
PO Box 888, Basseterre, St Kitts

ANTIGUA AND BARBUDA
Environmental Awareness Group
PO Box 2103, Long St, St John's
Primarily working through raising environmental awareness and advocacy.

MONTSERRAT
Montserrat National Trust
www.montserratnationaltrust.com/
PO Box 393, Olveston
Responsible for natural and cultural conservation. Undertaking some research on turtles.

GUADELOUPE
Union Régionale des Associations du Patrimoine et de l'Environnement de la Guadeloupe (URAPEG)
BP 273, 97110 Pointe a Pitre

DOMINICA
Dominica Conservation Association
PO Box 109, Roseau

Institute for Tropical Marine Ecology
www.itme.org
ITME Inc., PO Box 944, Roseau
Offers university level teaching, serves as a research station for visiting scientists, and is involved in community outreach.

MARTINIQUE
Association pour la Sauvegarde du Patrimoine Martiniquais (ASSAUPAMAR)
www.assaupamar.mq/
lm. Canavalia, Rés. du Square, Place d'Armes, 97232, Lamentin

Observatoire du Milieu Marin Martiniquais
7, avenue Condorcet, 97200 Fort de France
An NGO undertaking both reef research and public outreach.

ST LUCIA
Soufriere Marine Management Association
www.smma.org.lc/
The agency coordinating the management of the Soufriere Marine Management Area.

St Lucia National Trust
www.slunatrust.org/
PO Box 595, Castries
Manages terrestrial and coastal protected areas and strongly involved in environmental education and advocacy.

ST VINCENT
Mayreau Environmental Development Organization (MEDO)
Mayreau@caribsurf.com
Community-based organization on the island of Mayreau, west of Tobago Cays, working on awareness raising, education, and environmental management.

St Vincent National Trust
St Vincent National Trust, PO Box 752, Kingstown (Tel: 809 456 1060) or St Vincent National Trust, PO Box 1538, Kingstown
Works on conservation and protection of historic and natural resources.

GRENADA
Carriacou Environmental Committee
http://caribzones.com/cec.html
Raising public awareness and campaigning for better natural resource management.

National Trust and Historical Society
c/o Grenada National Museum, Young Street, St George's

BARBADOS
Barbados Marine Trust
www.barbadosmarinetrust.com/
Underwater Barbados, Bay Street, St Michael
A membership organization promoting environmentally and socially sustainable use of the marine areas of Barbados.

SOUTH AMERICA
COLOMBIA
Corporación para el Desarrollo Sostenible del Archipiélago de San Andrés, Providencia y Santa Catalina (CORALINA)
www.coralina.org
Aims to administer, protect, and restore the natural environment in the San Andrés and Providencia region. Administers Seaflower Biosphere Reserve.

Fundación Pro-Sierra Nevada de Santa Marta
www.prosierra.org/
Working on environmental protection in the Santa Marta area.

Instituto de Investigaciones Marinas y Costeras (INVEMAR)
www.invemar.org.co
Apartado aéreo 1016, Santa Marta
Marine research institute linked to the Ministry of the Environment, but also with some private funding, and in national and international partnerships.

Unidad Administrativa Especial del Sistema de Parques Nacionales Naturales
www.parquesnacionales.gov.co
Main government agency working on protected areas within the Environment Ministry.

VENEZUELA
Fundación para la Defensa de la Naturaleza (FUDENA)
www.fudena.org.ve
Av. Principal Los Cortijos de Lourdes c/2a, Caracas 1071-A
Major national environmental group undertaking advocacy, education, and research, and supporting management. Has a coastal program.

Instituto Nacional de Parques (INPARQUES)
www.inparques.gov.ve/
Government agency responsible for the management of protected areas.

ARUBA
Aruba Foundation for Nature & Parks (FANAPA)
Diamantbergweg 40, San Nicolas

NETHERLANDS ANTILLES (SOUTHERN GROUP)
Caribbean Marine Biological Foundation (CARMABI)
www.carmabi.org
PO Box 2090, Piscadera Baai, Curaçao
Undertakes marine and terrestrial research for the resource management of the Netherlands Antilles. Education program reaches more than 10 000 schoolchildren a year.

Curaçao Sea Aquarium
www.curacao-sea-aquarium.com/Education/
PO Box 3102, Curaçao
The Aquarium has an Education Department actively involved in teaching, both within the aquarium and through island-wide activities such as snorkeling clubs.

Reef Care Curaçao / Kuida Ref Korsou
PO Box 676, Willemstad, Curaçao
Aims to protect and conserve coral reefs, to perform research, and to raise public awareness. Ongoing activities include an educational project for underprivileged children, beach clean-up, the Reef Alarm Help Line, and the Coral/Sponge Spawning Program.

Coral Resource Management – Fundashon pa Bon Koral
www.coralresourcemanagement.org
Bara di Karta z/n, Bonaire

Working to support implementation of sound management of coral reefs both on Bonaire and internationally, using existing expertise to support training and management planning.

Sea Turtle Conservation Bonaire (STCB)
www.bonairenature.com/turtles/
PO Box 492, Kralendijk, Bonaire
Undertaking research and conservation activities, including turtle tracking, for marine turtles in Bonaire.

Stichting National Parken der NA Bonaire (STINAPA) (Foundation for the National Parks)
www.stinapa.org
Barkadera, s/n, PO Box 368, Bonaire
Manages the Bonaire Marine Park as well as other protected areas, heavily involved in public outreach, site management, conservation, and protection.

TRINIDAD AND TOBAGO
Buccoo Reef Trust
www.buccooreef.org/
TLH Building, Milford Road, Scarborough, Tobago
Seeks to protect reefs by education and outreach projects, with local people, and through ecotourism. Also developing the Tobago Marine Research Centre.

Environment Tobago
www.scsoft.de/et/et2.nsf
PO Box 503, Scarborough, Tobago
Raising environmental awareness, teaching schoolteachers, organizing clean-ups. Save Our Sea Turtles (SOS) Tobago is an affiliate.

Institute for Marine Affairs
www.ima.gov.tt/
Hilltop Lane, Chaguaramas, PO Box 3160, Carenage
Government research institution.

Trinidad and Tobago Field Naturalists' Club
www.wow.net/ttfnc/default.html
PO Box 642, Port of Spain
Organizes regular talks, meetings, and field visits.

INTERNATIONAL
Atlantic and Gulf Rapid Reef Assessment (AGRRA)
www.coral.noaa.gov/agra/
Atlantic and Gulf Rapid Reef Assessment, MGG-RSMAS, University of Miami, 4600 Rickenbacker Causeway, Miami, FL 33149, USA
International collaboration of scientists undertaking standardized surveys of coral reefs throughout the region aiming to determine the state of the coral reefs, and to make this information more widely available.

Caribbean Coastal Marine Productivity Program (CARICOMP)
www.ccdc.org.jm
CARICOMP Data Management Centre, c/o Center for Marine Sciences, University of the West Indies, Mona, Kingston 7, Jamaica
Cooperative research network of 29 marine laboratories, parks, and reserves, on 13 islands and in 9 mainland countries across the region. Each is undertaking standardized research on biodiversity and productivity in reefs, seagrasses, and mangroves.

Caribbean Conservation Association
www.ccanet.net
Chelford, The Garrison, St Michael, Barbados
Important regional organization whose members are other organizations and governments, working to promote conservation across the region. Its Coastal and Marine Management Programme has worked, among other things, on a Coral Reef Education Project for Schools.

Caribbean Conservation Corporation
www.cccturtle.org
4424 NW 13th Street, Suite #A1, Gainsville, Florida 32609, USA
Focuses on sea turtle conservation, including satellite tracking.

Caribbean Natural Resource Institute (CANARI)
www.canari.org/
Fernandes Industrial Centre, Administrative Building, Eastern Main Road, Laventille, Trinidad
Technical and research organization, which analyzes and promotes

participatory management of natural resources in the islands of the Caribbean, through research and capacity building.

Conservation International
www.conservation.org
1919 M Street NW, Suite 600, Washington DC 20036, USA
A global field-based organization taking part in science and conservation projects worldwide, much of its work is focused on hotspots where biodiversity is particularly rich, and these include both Mesoamerica and the Caribbean islands.

Environmental Defense
www.environmentaldefense.org
14630 SW 144 Terrace, Miami FL 33186, USA
A membership organization that links science, law, and economics to create solutions to environmental problems. Has a marine program.

Global Coral Reef Monitoring Network (GCRMN)
www.gcrmn.org
Australian Institute of Marine Science, PMB 3, Townsville MC, Townsville 4810, Queensland, Australia
An operation unit of ICRI seeking to support and help coordinate global coral reef monitoring at the levels of communities, managers, and the scientific community.

International Coral Reef Action Network (ICRAN)
www.icran.org
c/o UNEP-WCMC, 219 Huntingdon Road, Cambridge CB3 0DL, UK
A partnership of organizations (many listed here) supporting coral reef monitoring and assessment, information dissemination and communication, and reef management.

International Coral Reef Initiative
www.icriforum.org
c/o UNEP-WCMC, 219 Huntingdon Road, Cambridge CB3 0DL, UK
A partnership among governments, international organizations, and NGOs to improve reef conservation, helping to improve cooperation and coordination between governments, donors, and other sectors.

International Society for Reef Studies
www.fit.edu/isrs/
The leading academic coral reef
organization, offers grants,
organizes major scientific meetings, and
publishes *Coral Reefs* and *Reef
Encounter*.

Island Resources Foundation
www.irf.org/
*Library and Island Systems Centre,
123 Main Street, Box 3097, Road Town,
Tortola, BVI*
A private, non-profit research and
education organization dedicated to
solving the environmental problems of
development in small tropical islands.
Important networking and internet
information provider.

IUCN–The World Conservation Union
www.iucn.org/themes/marine/
*IUCN–The World Conservation Union,
Global Marine Program, 1630
Connecticut Avenue NW, Suite 300,
Washington DC 20009-1053*
A unique organization whose members
are conservation organizations,
government agencies, and states. It has
offices in Costa Rica and Ecuador, but
most of its Caribbean work is carried out
through its Global Marine Program in
Washington DC.

Marine Aquarium Council
www.aquariumcouncil.org
*923 Nu'uanu Avenue, Honolulu,
Hawaii, 96817 USA*
Working with the marine aquarium
industry and hobbyists, public aquariums,
and conservation organizations, it has set
standards and established a certification
scheme for the global trade in marine
aquarium organisms.

Project AWARE Foundation
www.projectaware.org/
*30151 Tomas Street, Rancho Santa
Margarita CA 92688-2125, USA*
Linked to PADI (Professional Association
of Diving Instructors) the foundation runs
public awareness and educational
campaigns and provides support for a
broad range of projects.

Project Seahorse
www.projectseahorse.org/
*University of British Columbia, Fisheries
Centre, 2259 Lower Mall, Vancouver, BC
V6T 1Z4, Canada*
A professional team, working to promote
seahorse conservation through research,
education, and active engagement in
policy at local and national levels.

Reef Check
www.reefcheck.org
*Reef Check Headquarters, Institute of the
Environment, 1362 Hershey Hall, Box
951496, University of California at Los
Angeles, Los Angeles CA 90095-1496,
USA*
The largest global reef monitoring
organization, using community groups
and volunteers to undertake standardized
reef research, with coordinators in over
20 Caribbean countries. Helps to
network, encourages education and
conservation, and provides scientific
information from sites across the globe.

**Reef Environmental Education
Foundation (REEF)**
www.reef.org
*REEF, 98300 Overseas Hwy, Key Largo
FL 33037, USA*
An organization of recreational divers
who regularly conduct simple fish
biodiversity and abundance surveys.
Making use of volunteers, more than
65 000 surveys have now been
undertaken. The divers learn about the
reef environment and their findings have
been used in a number of scientific
publications.

Reef Relief
www.reefrelief.org/
PO Box 430, Key West FL 33041, USA
An organization based in Florida and the
Bahamas, involved in educational,
scientific, and campaigning projects,
including work with schools, installing
mooring buoys, and reef monitoring.

**Small Island Developing States
Network (SIDSnet)**
www.sidsnet.org
*SIDSnet, United Nations Division for
Sustainable Development, United Nations
Department of Economic and Social
Affairs, 2 UN Plaza, DC2 – 2020, New
York NY 10017, USA*

A networking organization joining 43
small island developing states (SIDS), to
support sustainable development
through information provision and
networking.

The Coral Reef Alliance (CORAL)
www.coralreefalliance.org/
*417 Montgomery Street, Suite 205, San
Francisco CA 94104, USA*
A membership-based organization based
in the USA but working in the western
Pacific and Caribbean. A key educator on
coral reefs, working with marine park
managers, businesses, and communities.
Also hosting the International Coral Reef
Information Network (ICRIN),
www.coralreef.org, a valuable
information resource.

The Nature Conservancy (TNC)
http://nature.org/
*4245 North Fairfax Drive, Suite 100,
Arlington VA 22203-1606, USA*
A global environmental group working at
local levels through national and state
offices and with partners across the
region. The marine initiative is working
through research, advocacy, and
outreach, particularly focusing on coral
reefs and marine protected areas.

The Ocean Conservancy
www.oceanconservancy.org
*1725 DeSales Street, NW Suite 600,
Washington DC 20036, USA*
An active campaigning organization,
dedicated to improving the health of the
oceans through advocacy, research, and
public education, in the USA and
internationally.

The World Fish Center and ReefBase
www.worldfishcenter.org and
www.reefbase.org
*PO Box 500, GPO, 10670, Penang,
Malaysia*
The World Fish Center is working on
tackling poverty by improving the state
and productivity of aquatic resources. It
has built up the highly detailed and
valuable database on coral reefs,
ReefBase, and was responsible for the
excellent fish database, FishBase.

Turtles in the Caribbean Overseas Territories
www.seaturtle.org/mtrg/projects/
tcot/
Working on turtles, mostly hawksbills in the British Overseas Territories, especially BVI, Caymans and Montserrat.

United Nations Environment Programme (UNEP)
www.unep.org
United Nations Avenue, Gigiri, PO Box 30552, Nairobi, Kenya
The leading United Nations organization concerned with the environment and promoting implementation of international agreements. Much of its coral reef work is conducted at the UNEP World Conservation Monitoring Centre and through its regional offices, including the Caribbean Environment Programme.

UNEP Caribbean Environment Programme (UNEP-CEP)
www.cep.unep.org/
UNEP Caribbean Environment Programme, Regional Co-ordinating Unit, 14-20 Port Royal Street, Kingston, Jamaica
Provides the secretariat for the only regional environmental treaty (the Cartagena Convention), including its various protocols such as the Specially Protected Areas and Wildlife Protocol. Provides assistance to all countries, helping to strengthen environmental policy and institutions, and to coordinate international cooperation. Disseminates information including CEP studies implemented with regional partners, organizes meetings and training workshops.

UNEP World Conservation Monitoring Centre (UNEP-WCMC)
www.unep-wcmc.org
UNEP-WCMC, 219 Huntingdon Road, Cambridge CB3 0DL, UK
The biodiversity information and assessment center for the United Nations, working in numerous public and private partnerships to provide accurate information to support policy and action. The global leader in habitat mapping (notably coral reefs, seagrasses, and mangrove forests), and in maintaining databases on protected areas and threatened species.

Wider Caribbean Sea Turtle Conservation Network
www.widecast.org/
Coordinates research and raising awareness of marine turtle conservation across the region.

World Resources Institute
www.wri.org
10 G Street, Suite 800, Washington DC 20002, USA
A key research organization looking at global and regional threats to coral reefs, and coordinating a collaborative research exercise into Reefs at Risk in the Caribbean which combines research, education, and awareness raising (available online).

World Wide Fund for Nature (WWF)
www.panda.org
Avenue du Mont Blanc, 1196 Gland, Switzerland
A global membership organization, with international, regional, and national offices campaigning for conservation, raising awareness, and supporting research and practical conservation initiatives. Runs an Endangered Seas Programme; works with the Marine Stewardship Council on certification of sustainable marine fisheries; actively supports conservation activities in the Mesoamerican Reef. Global office in Switzerland, and regional offices in the USA (World Wildlife Fund), Mexico, Costa Rica, and Colombia.

SEAFOOD GUIDES
The following organizations offer guidance on seafood which may be eaten in different places without contributing to overfishing. To date they have not provided a detailed coverage of coral reef areas, but this may change.

Seafood Choices Alliance
www.seafoodchoices.com/

Audubon Society
http://seafood.audubon.org/

Monterey Bay Aquarium
www.mbayaq.org/cr/seafoodwatch.asp

Environmental Defense
www.environmentaldefense.org

Marine Conservation Society
www.mcsuk.org/

VOLUNTEER ORGANIZATIONS
Many environmental organizations listed above welcome input from volunteers. In addition there are many which operate in several countries and invite paying volunteers to join expeditions where they can support research into coral reefs.

Coral Cay Conservation
www.coralcay.org
The Tower, 13th Floor, 125 High Street, London SW19 2JG, UK
Worked in Belize for many years, is currently in Honduras, and has undertaken recent research visits to Tobago Keys (St Vincent) and Sian Ka'an (Mexico).

Earthwatch
www.earthwatch.org
680 Mt Auburn Street, Watertown MA 02272, USA
Volunteers join individual scientists on targeted research projects. Recent and ongoing work in the Caribbean includes research into coral reproduction, coral disease, leatherback turtles, manatees, whales, dolphins, and damselfish.

Frontier
www.frontier.ac.uk/
Frontier, 50-52 Rivington Street, London EC2A 3QP, UK
Undertaking expeditions in Nicaragua.

Greenforce
www.greenforce.org/
11-15 Betterton Street, Covent Garden, London WC2H 9BP, UK
Volunteer expeditions are involved in surveying reefs in and around the new marine parks on Andros Island, and working with local communities in training and education programs.

Operation Wallacea
www.opwall.com/default.htm
Operation Wallacea, Hope House, Old Bolingbroke, Spilsby, Lincolnshire PE23 4EX, UK
Running expeditions in various countries including marine research expeditions to Cayos Cochinos, Honduras.

FURTHER READING AND SOURCES

FURTHER READING
IDENTIFICATION GUIDES

A Field Guide to Coral Reefs: Caribbean and Florida. EH Kaplan (1999). Houghton Mifflin Co.

Guide to Marine Life: Caribbean, Bahamas, Florida. M Snyderman, C Wiseman (1996). Aqua Press Publications, New York.

Marine Plants of the Caribbean: A Field Guide from Florida to Brazil. D Scullion Littler *et al* (1992). Smithsonian Institution Press.

National Audubon Society Field Guide to Tropical Marine Fishes of the Caribbean, the Gulf of Mexico, Florida, the Bahamas, and Bermuda. C Lavett Smith (1997). Chanticleer Press, New York.

Reef Coral Identification: Florida Caribbean Bahamas. P Humann, N Deloach (2001). New World Publications.

Reef Creature Identification: Florida Caribbean Bahamas. P Humann, N Deloach (2001). New World Publications.

Reef Fish Identification: Florida Caribbean Bahamas. P Humann, N Deloach (2002). New World Publications.

Reef Fishes, Corals, and Invertebrates of the Caribbean. E Wood, L Wood (2000).

Snorkeling Guide to Marine Life: Florida Caribbean Bahamas. P Humann, N Deloach (1995). New World Publications.

TRAVEL GUIDES

A number of travel guides provide detailed information on diving in single countries or groups of countries. These are not listed here, but three series are considered to be of excellent quality, providing information for divers and snorkelers down to the level of individual dive locations.

Pisces Diving and Snorkelling Guides. Published by Lonely Planet. Currently available for: Bahamas, Belize, Bermuda, Bonaire, British Virgin Islands, Cayman Islands, Cozumel, Cuba, Dominica, Florida Keys, Honduras' Bay Islands, Puerto Rico, Trinidad & Tobago, Turks & Caicos Islands, US Virgin Islands.

The Dive Sites of the World Series. Published by New Holland Publishers. Currently available for: Aruba, Bonaire and Curaçao, Bahamas, Philippines, Cayman Islands, Cozumel and the Yucatán, Virgin Islands.

The Complete Diving Guides. C Ryan, B Savage, Complete Dive Guide Publishers. A three-volume series covering all of the islands from Puerto Rico and the Virgin Islands to Tobago.

GENERAL

National Geographic Atlas of the Ocean. SA Earle (2001). National Geographic Society.

Reef Fish Behavior: Florida Caribbean Bahamas. N Deloach, P Humann (1999). New World Publications.

Sea Change: A Message of the Oceans. SA Earle (1996). Ballantine Books.

The Enchanted Braid: Coming to Terms with Nature on the Coral Reef. OG Davidson (1998). John Wiley and Sons.

TECHNICAL

Caribbean Marine Protected Areas: Practical Approaches to Achieve Economic and Conservation Goals. RS Appeldorn *et al* (2003). Gulf and Caribbean Research Vol 14 No. 2. University of Southern Mississippi.

CARICOMP – Caribbean Coral Reef, Seagrass and Mangrove Sites. UNESCO (1998). Coastal Region and Small Island Papers 3, UNESCO, Paris.

Coral Reef Fishes: Dynamics and Diversity in a Complex Ecosystem. P Sale (2002). Academic Press.

Fully Protected Marine Reserves: A Guide. CM Roberts, JP Hawkins (2000). WWF Endangered Seas Campaign, Washington DC.

Latin American Coral Reefs. Cortés J (ed) (2003). Elsevier Science.

Life and Death of Coral Reefs. C Birkeland (ed) (1997). Chapman and Hall, New York.

Seas at the Millennium: An Environmental Evaluation. 3 vols. C Sheppard (ed) (2000). Elsevier Science.

Status of Coral Reefs in the Western Atlantic. JC Lang (ed) (2003). Atoll Research Bulletin No 496. Smithsonian Institution, Washington DC.

Status of Coral Reefs of the World: 2002. C Wilkinson (2002). Australian Institute of Marine Science, Townsville.

The Fishery Effects of Marine Reserves and Fishery Closures. FR Gell, CM Roberts. WWF-US, Washington DC.

World Atlas of Coral Reefs. MD Spalding *et al* (2001). University of California Press, Berkeley.

World Atlas of Seagrasses. EP Green, FT Short (eds) (2003). University of California Press, Berkeley.

SOURCES

In compiling this book the author used numerous references, including those listed above, but also many detailed scientific papers, other reference books, and the internet.

Under the individual countries, coral reef areas are taken from the *World Atlas of Coral Reefs.* Temperature and rainfall information are taken from www.weatherbase.com whenever possible. They are taken from the capital, unless specified (sometimes a coastal town near the coral reefs is more useful). For most places the range represents the mean high and mean low temperature for the specified month/s and gives an idea of the sort of variation that might be expected.

Conservation International provided valuable information in support of a paper prepared by the author and Phil Kramer for its Defying Ocean's End meeting. These included tourism statistics, mostly derived from the World Tourism Organization. In almost all cases they are annual statistics relating to the year 2000. They also included the number of islands statistics calculated for the island nations. For the purposes of this calculation "large" islands were any greater than 1 square kilometer. These figures may be a little different from other sources, but the value of presenting our own statistics is that they have been accurately counted from a standardized map.

INDEX